THE SPACE TELESCOPE

OTHER BOOKS BY THE AUTHORS

George Field
  The Redshift Controversy (with Halton Arp and John
    Bahcall)
  The Dusty Universe (ed. with Alastair G. W. Cameron)
  Cosmic Evolution: An Introduction to Astronomy (with
    Gerrit Verschuur and Cyril Ponnamperuma)
  The Invisible Universe (with Eric Chaisson)

Donald Goldsmith
  From the Black Hole to the Infinite Universe (with
    Donald Levy)
  The Universe
  Scientists Confront Velikovsky (ed.)
  What Is a Star?
  The Search for Life in the Universe (with Tobias
    Owen)
  The Quest for Extraterrestrial Life (ed.)
  The Evolving Universe
  Cosmic Horizons (with Robert Wagoner)
  *Nemesis: The Death-Star and Other Theories of Mass
    Extinction*
  Supernova!

THE SPACE TELESCOPE

DR. GEORGE FIELD
and DR. DONALD GOLDSMITH

CB

CONTEMPORARY
BOOKS

CHICAGO · NEW YORK

Library of Congress Cataloging-in-Publication Data

Field, George B., 1929–
    The space telescope : eyes above the atmosphere / George
Field and Donald Goldsmith.
       p.   cm.
    Includes bibliographic references.
    ISBN 0-8092-4495-0
    1. Hubble Space Telescope.   I. Goldsmith, Donald.
  II. Title.
  QB500.268.F54   1989
  522'.2919—dc20   89-22375
                                                    CIP

Published by Contemporary Books, Inc.
180 North Michigan Avenue, Chicago, Illinois 60601
Manufactured in the United States of America
International Standard Book Number: 0-8092-4495-0

Published simultaneously in Canada by Beaverbooks, Ltd.
195 Allstate Parkway, Valleywood Business Park
Markham, Ontario L3R 4T8 Canada

To Susan Field and Rachel Goldsmith

CONTENTS

ACKNOWLEDGMENTS

IN WRITING THIS book, we have received tremendous assistance from a host of astronomers and others who gave generously of their time and energy. We are grateful for this aid, and hope that it has helped us to tell the story of the Space Telescope correctly. We must stress that any errors are the fault of the authors.

Our special thanks go to John Bahcall, Eric Chaisson, Charles O'Dell, and Lyman Spitzer for taking great pains with the manuscript. We also wish to thank Ken Brecher, Paul Goldsmith, Steve Maran, Irwin Shapiro, and most of all Riccardo Giacconi for supplying helpful information. Nancy Lee Snyder and Diane Steiner lent their skills to preparing the manuscript and finding the illustrations, and Marjorie Baird Garlin was constantly patient in drawing the line art.

May all of these people take pleasure in what the Space Telescope will achieve.

1

LAUNCH

THE PLACE IS CAPE Canaveral, Florida; the year is 1990. The United States, after nearly half a decade without a significant launch of a space exploration vehicle, has revitalized the space program with the *Magellan* mission to Venus and the *Galileo* probe to Jupiter. Now it is time to take the lead in the discovery of the cosmos. The Hubble Space Telescope, by far the mightiest scientific instrument ever sent above the Earth's enshrouding atmosphere, has waited more than four years for this moment, preserved in the world's largest, most expensive clean room.

The giant telescope, large as a trailer truck but able to point with exquisite precision toward any location on the sky, has flown on a C5A aircraft, sealed in a special container to maintain its environment, from Lockheed's facility in Sunnyvale, California, to the Kennedy Space Flight Center in Florida. There a host of technicians have assured its readiness for the near-vacuum of space, 375 miles above the Earth's surface, to which the telescope will be borne by the Space Shuttle, once again certified for safe flight.

The Hubble Space Telescope (Space Telescope for short) is a crucial payload for the Space Shuttle—the heaviest civilian satellite the shuttle has ever launched—and the most important scientific instrument ever prepared for orbit above Earth's atmosphere. On its successful launch has come to rest the success or failure—at least for the immediate future—of the United States' effort to open the

1

universe to human understanding by sending instruments into space.

Although the United States has been the driving force behind the Space Telescope, the telescope project involves collaboration with the European Space Agency, which embraces fourteen European countries, including France, England, and West Germany. Hence the Space Telescope is an international effort, and the scientists gathered to witness the launch include not only a host of Americans but also the outstanding figures of the European space science community. But the Americans predominate: the Space Telescope operates on commands sent from the Space Telescope Science Institute at the Johns Hopkins University in Baltimore and from NASA's Goddard Space Flight Center in Greenbelt, Maryland. Any failure of the Space Telescope will be laid at those doorsteps. For that reason, the most intense figure at the telescope's launch, and the pivotal man in the effort to operate the Space Telescope, will be Riccardo Giacconi, the Italian-born scientist who heads the Space Telescope Science Institute (Figure 1).

Figure 1: Riccardo Giacconi (1931–   ), Director of the Space Telescope Science Institute. *The Space Telescope Science Institute*

Giacconi radiates energy like a star. Heavyset, dark-haired, of medium height, in his late fifties, he is a hard-driving, tough-talking, nervously moving, completely dedicated scientist. On Giacconi rests the responsibility for assuring that the Space Telescope does *science*; although NASA is committed to the successful launch and operation of the Space Telescope, it is Giacconi's job to assure not only that data emerge from the instrument but also that these data represent useful, important scientific information.

The launch of the Space Telescope will culminate a decade of struggle to design, build, and test it; this decade in turn caps the forty years of U.S. space exploration that followed the end of the Second World War. Giacconi became the director of the Space Telescope Science Institute in 1981. Prior to that, he had worked for American Science & Engineering, a corporation in Cambridge, Massachusetts, largely staffed by technicians and scientists, and later served as associate director of the Harvard-Smithsonian Center for Astrophysics in Cambridge.

At AS&E, Giacconi quickly became known as a man who made things happen. Smoking cigarette after cigarette, speaking loudly in his accented English, Giacconi might have been a figure of fun were it not for his complete mastery of both the engineering and scientific aspects of the satellites on which he worked—and for his ability to dominate a discussion among equally heavy hitters.

Of all the scientists at the Space Telescope launch, Lyman Spitzer, Jr. (Figure 2) will be the most venerated and significant figure of the postwar period. Spitzer comes from a banking family in Toledo, Ohio (where one can still enter the Spitzer building), but he chose a career in science more than fifty years ago. During the mid-1930s, Spitzer studied astrophysics at Cambridge University in England with Arthur Stanley Eddington, then the world's premier astrophysicist, and wrote his Ph.D. dissertation at Princeton University under the supervision of Henry Norris Russell, the outstanding astrophysicist of the United States. Rus-

Figure 2: Lyman Spitzer, Jr. (1914–   ), who first seriously proposed placing an automated telescope in orbit above the atmosphere.

sell had made Princeton this country's astrophysics center, and Spitzer has now spent more than forty years on the Princeton faculty. He once said that he would consider leaving Princeton only if by staying he would prevent the Space Telescope from becoming a reality.

The Space Telescope traces its roots directly to Lyman Spitzer. After scientific service during World War II (during which, Spitzer wryly notes, he studied undersea warfare on the upper floors of the Empire State Building), Spitzer in 1946 wrote a report for the RAND Corporation, which was engaged in research for the U.S. Air Force. This report examined the feasibility—more than a decade before the launch of the first artificial Earth satellite—of placing satellites into Earth orbit in order to observe radiation from space that cannot penetrate our planet's atmosphere. Spitzer suggested orbiting a large astronomical telescope above the blurring, changing atmosphere of Earth, which prevents us from obtaining a perfectly clear view of the heavens, no matter how perfect our Earthbound instruments may be.

When this report appeared, Spitzer was a young professor at Yale University. A year later, Princeton called him home to succeed Russell as head of the Princeton Observatory. There Spitzer continued to pursue his goal of launching astronomical satellites. For relaxation, he began to climb mountains. A tall, lanky man, Spitzer has stood atop more peaks than most of us can name. At age sixty-one, he scaled the tower of the Graduate College at Princeton—and was arrested for his efforts, though soon released. Now in his mid-seventies, Spitzer is driven by the desire to see the Space Telescope fly, to realize a dream he had more than forty years ago.

During the 1950s, astronomical observations at Princeton concentrated on Project Stratoscope, an effort to use high-altitude balloons to carry automated telescopes to altitudes twenty miles above the Earth, where the thinness of the atmosphere allows sharper pictures than are possible from the ground (Figure 3). Stratoscope took fine pictures of the sun, the planets, and other objects, but the perceived importance of the project, and plans for eventual satellite observations, were overtaken by events.

In October 1957, the Soviet Union launched *Sputnik I*, the first human-made satellite of Earth, and followed this a month later with the much heavier *Sputnik II*. Not only had the Russians beaten America into space, they had also demonstrated a much greater launch capability. It is no exaggeration to state that panic temporarily engulfed the segments of the public, and especially of the government, that saw rocket technology as the key to military might.

The next few years brought an all-out effort in the United States to regain equality, if not superiority, in rocket vehicles and satellite launches. The United States

Figure 3: An automated telescope to be lofted to high altitudes by a balloon as part of Princeton University's Project Stratoscope. *Princeton University Observatory*

struggled mightily to launch its first satellite, in January 1958, and to produce a new generation of rockets. During the following few years, the question of whether or not a "missile gap" existed between the United States and the Soviet Union colored public perception of rocket development, and the United States embarked on a host of new rocket programs, some of which would be canceled a few years later. In 1962, President Kennedy announced that the

United States would place men on the moon by the end of the decade and committed the country to an ambitious space exploration program that indeed resulted in the first men on the moon in 1969.

To the public, and largely to the government as well, space exploration meant *manned* space exploration. Automated space probes seemed all very well in their way, but nothing could command attention like a space capsule with a crew. For decades, scientists like Spitzer and Giacconi have argued for a more reasonable balance between manned and unmanned space probes. To support a human being in space requires expenditures for space, air, food, disposal systems—and above all safety—that an automated probe does not need. But, as Spitzer and Giacconi know better than anyone, the most cogent arguments must confront the all-too-human feeling that if it's not alive, it's not so important. In the end, after years of struggle, they and other scientists and engineers obtained approval for a series of automated probes for strictly scientific purposes.

So the United States prepared and launched a series of lunar and planetary explorers—Rangers, Pioneers, Voyagers, Vikings—that unlocked old mysteries associated with the moon and planets and uncovered a host of new facts ripe for further speculation and investigation. With the launching of *Voyager 2* in 1977, the U.S. planetary exploration program entered a limbo that has persisted for more than a decade, to resume in 1989 with the *Magellan* and *Galileo* spacecraft, the first launched to map Venus's surface by radar, the second to travel via a roundabout route past Venus and twice back past the Earth to reach the planet Jupiter early in 1995. But in addition to the probes sent into the solar system for close-up examination of other worlds there, a series of satellites in orbit *around* the Earth reshaped our thinking about the far greater cosmos, the vast spaces that lie far beyond the sun, its planets, and their satellites.

How could spacecraft orbiting only a few hundred miles above the Earth add so significantly to our store of knowledge? The answer lies in our life-giving, protective atmosphere, the blanket of air that perpetually wraps our planet in a not-quite-invisible shroud. Our atmosphere lets pass most of the visible light—the light that our eyes detect— although it does so only with some blurring of the light, the effect that makes stars "twinkle." But other types of light from the stars, in particular ultraviolet light, are absorbed by the gases of our atmosphere. This fact allows us to live, for the sun produces enough ultraviolet light to kill us, were it not blocked by our atmosphere. Not surprisingly, we and other animals on Earth have evolved eyes that do not detect ultraviolet light, though our laboratory skills are now highly developed in this pursuit. But the best detectors remain useless if ultraviolet light from the stars cannot reach them.

Hence the need for space satellites high enough in orbit to remain above the layer of ozone and other absorbing molecules in our atmosphere, capable of observing the cosmos as it really is, rich in ultraviolet and all the other types of light that cosmic objects produce. Lyman Spitzer saw this need clearly in 1946, when he proposed a series of satellites capable of viewing the universe in all types of light—ultraviolet, infrared, microwave, radio, gamma-ray, x-ray. But from the dream to the reality took many steps and almost as many setbacks. Before the late 1970s, there were significant steps toward Spitzer's goal—which had become the goal of most astronomers—comprising a series of four astronomical satellites called the Orbiting Astronomical Observatories (OAOs).

The four OAOs were conceived and designed primarily by astronomers at the Smithsonian Astrophysical Observatory, Princeton University, the Goddard Space Flight Center, and the University of Wisconsin, working in collaboration with scientists and engineers at NASA and at the contracting firms that built the satellites. All four of the satellites aimed to study the sky in ultraviolet light and thus to obtain the first clear look at the universe in this type of radiation that never reaches the Earth. *OAO-A* had

Figure 4: The *OAO-3/Copernicus* satellite, which carried out observations of the cosmos in ultraviolet radiation during the 1970s, here being assembled for flight. *NASA*

a battery malfunction and never performed satisfactorily. *OAO-B* experienced a rocket failure and had a short ride into the Indian Ocean. But *OAO-A2* (designated *OAO-2* after it attained orbit) and *OAO-C* (designated *OAO-3* after it attained orbit, and later renamed *Copernicus* to commemorate the famous astronomer's 500th birthday) were clear successes (Figure 4). They lifted the veil beneath which astronomers—and the rest of us—had lived since time immemorial; they gave us our first broad look at the heavens in ultraviolet light; they created a new breed of astronomers, "ultraviolet astronomers"; and they paved the way for the Space Telescope, which would observe the cosmos in both visible and ultraviolet light.

And what had Riccardo Giacconi been up to during the 1970s, while Lyman Spitzer and other astronomers and engineers were bringing the OAO satellites to fruition? Giacconi had made his mark in another domain of astronomy, x-ray astronomy. Like ultraviolet, x-rays represent another type of radiation, a type completely invisible to human eyes. X-rays, more penetrating than ultraviolet or visible light, pass easily through human flesh (not so easily through human bones) and can be detected by special photographic film. For this reason, we all know their usefulness: a source of x-rays, in combination with x-ray film, can register pictures of what lies inside us. And we all should know the danger of x-rays: prolonged exposure to them can cause cancer and harm to our bodies. Here again, our atmosphere protects us, since it allows no x-rays from space to pass. We can therefore salute our protective blanket of air—even as astronomers dream of placing their instruments outside it.

Giacconi helped to lead the way. X-ray astronomy began with instruments carried by rockets to heights of many dozen miles, high enough to avoid most of the atmospheric blockage of the x-rays arriving from deep space. The first detections of x-rays from the sun were made from captured German V-2 rockets launched from New Mexico. Then, led by Herbert Friedman, a team from the Naval Research Laboratory followed on this work to perform pioneering observations of various x-ray-emitting objects in the sky.

But the first true x-ray *survey* of the sky—a look in every direction to find out which objects are emitting x-rays—remained only a plan so long as only rocket-borne observations could be made. Rocket flights last just a few minutes, not long enough to engage in survey observations. For that you need an x-ray-detecting satellite, and the man for the survey instrument was Riccardo Giacconi.

Giacconi came to the United States from Italy during the mid-1950s and soon became deeply involved in x-ray astronomy. In 1962 he led a team that launched a rocket-borne detector that found the first x-ray-emitting object outside the solar system. During the following fifteen years, Giacconi became the key figure of the first two x-ray satellites, the *Uhuru* satellite, launched in 1970, and the *Einstein* satellite, launched in 1978.

Uhuru, which means "freedom" in Swahili, owes its name to the fact that the satellite was launched from a floating platform off the coast of Kenya in East Africa. This position, directly on the Earth's equator, allowed the satellite to reach an orbit in which it circled the Earth always directly above the equator—the ideal orbit for a survey satellite, for it allows the satellite to see every bit of the sky (Figure 5). *Uhuru* made the first x-ray survey of the sky, discovering hundreds of x-ray sources totally unknown to astronomers.

Figure 5: An artist's conception of the *Uhuru* satellite, launched into synchronous orbit from the equator near the coast of Kenya. *NASA*

The *Einstein* satellite, named to honor Albert Einstein's birth centennial (1979), could not only detect x-ray sources but could also image them. That is, it could obtain x-ray pictures of the objects that emit x-rays, recording the patterns of x-rays that each such object emits. Among its other triumphs, the *Einstein* satellite showed that virtually every star is an "x-ray star," a source of some x-rays, although most stars, like our sun, shine only weakly in x-rays.

By the end of the 1970s, Giacconi had demonstrated that he could make a complex satellite project a success. Although his sometimes abrasive manner had struck sparks on occasion, Giacconi was a natural choice to head the Space Telescope Science Institute, which NASA chose to locate in Baltimore at the Homewood Campus of the Johns Hopkins University. The holder of the directorship at the institute was destined to be—and will eventually become— the key figure in our attempts to observe the universe from beyond the Earth's atmosphere.

Data from the Space Telescope will flow simultaneously to NASA's Goddard Space Flight Center and to the Space Telescope Science Institute, but the latter research center is charged with the task of processing and distributing the data to astronomers who know how to interpret such information. For that reason, a host of astronomers look to the Space Telescope Science Institute, and thus to Giacconi, to make the Space Telescope work in the fullest sense: to see that it operates properly as a scientific instrument, and to see that their observations are made and the data transmitted correctly back to the institute and on to the individuals who seek to use it. For all of this, Giacconi is ready. All that he needs is to have the Space Telescope launched into orbit, and that is the responsibility of NASA.

Carrying the Space Telescope into orbit requires the Space Shuttle, a fact that stems from a decision made fifteen years ago, when NASA decided to mate the proposed "Large Space Telescope," as it was then called, with the space transportation system of the future. No one then foresaw the difficulties in making the Space Shuttle work properly, let alone the three-year hiatus that would follow the *Challenger* disaster of January 1986. But the design problems have been overcome, the disaster surmounted. Once again the Space Shuttle is in operation, with the new, safer design of the solid-fuel rocket booster that had exploded on that fateful January morning.

And so the Space Shuttle will be readied at Cape Canaveral, steaming as always from its enormous external tank containing 790 tons of liquid fuel: 143,000 gallons of liquid oxygen in the upper tank and 383,000 gallons of liquid hydrogen in the lower. So that the oxygen and hydrogen will remain in the liquid state, the shuttle needs an elaborate cooling system to keep the oxygen at −297 degrees Fahrenheit and the hydrogen at −423 degrees, just 50 degrees above absolute zero. By themselves, the hydrogen and oxygen remain inert; but let them combine inside the Space Shuttle's engines, and you have as explosive a mixture as any rocket could ask for. On either side of the main external tank stand the solid-rocket boosters, whose aluminum-based fuel burns steadily but uncontrollably. Engineers can, with good design, turn off the liquid-fuel motor upon command, by halting the flow of liquid oxygen and hydrogen into the chamber where they meet and ignite. But once you ignite the "solids," as the Cape Canaveral engineers call them, you must be prepared to let them burn completely, for there is no way to shut them off. So if you are launching a space shuttle, you had better be sure that your design and operations plan is a good one; otherwise, disaster stares you in the face.

For the Space Telescope, astronomers count on success. The Space Shuttle's main engines, using fuel from the external tank, must ignite; then the solid-rocket boosters must ignite within a second of each other to avoid an asymmetrical thrust that would destroy the spacecraft. Belching steam from its hindquarters, packed with the largest payload it can carry, the Space Shuttle will rise slowly and magnificently from its launch tower. Twelve seconds after lift-off, before the enormous, deep-throated roar of the engines reaches the crowd of spectators nine miles away, the shuttle must roll onto its back, the desired position for further acceleration, and rise at full power.

Soaring more and more rapidly, the spacecraft reaches an altitude of five miles before the first minute of flight has elapsed. After two minutes, with the spacecraft twenty miles high, the solid-rocket boosters, each as high as a fifteen-story building but now devoid of fuel, must separate from the shuttle and tumble toward the sea, deploying 100-foot-wide parachutes to deaden their impact with the ocean. The Space Shuttle is fifty miles high, then a hundred. Onboard computers, backed by computers on the ground, command the shuttle's own liquid-fuel motor to push the spacecraft toward its desired orbit, 375 miles above the Earth, an orbit that carries the shuttle from twenty-eight degrees north latitude (the latitude of the Kennedy Space Flight Center) to twenty-eight degrees south latitude. Then the external liquid-fuel tank, likewise exhausted of fuel, must fall away.

At this point, the shuttle must ignite its main engine once more to achieve a circular orbit 375 miles above the terrestrial globe. Then the shuttle can use its momentum to coast around and around the Earth, taking ninety-six minutes for each orbit. The Earth's gravity will pull the spacecraft downward, while the spacecraft's momentum—its tendency to keep moving in the same direction and at the same speed—will impel the shuttle toward a straight-line trajectory. The battle of the two effects, gravity and momentum, will keep the shuttle in a nearly circular orbit. Turn off the Earth's gravity, and the shuttle would sail off into space. Turn off its momentum—slow the shuttle down—and it would fall back to Earth. But by achieving the perfect balance between gravity and momentum, the shuttle can ride around and around the Earth. As Isaac Newton had anticipated in his book *The System of the World*, the shuttle will fall around the planet, like a ball shot from such a powerful cannon that, instead of falling only partway around the curved Earth, it will fall all the way around the Earth, forever.

Once the shuttle achieves orbit, the astronauts and everything else on board begin to experience what is misnamed "weightlessness" or sometimes, even more incorrectly, "microgravity." In fact, gravity works just as well on the Space Shuttle as on the Earth's surface, once we allow for the fact that the shuttle is a few hundred miles farther from the Earth's center than we are. The shuttle "weighs" in space just about what it does on the ground (not counting the fuel that it burns to achieve orbit). What happens in orbit is not that gravity turns off but that an additional factor appears—the constant fall of the spacecraft around the Earth. Jump from a high mountain, and you too will feel "weightless"; more correctly, nothing will push back on you, as the floor does when your weight presses upon it.

So the astronauts may say that they seem "weightless," when they might better say that both they and their microenvironment have begun an ongoing state of free-fall around the world. Free-fall takes some getting used to, but the astronauts have learned how to function rather well while falling. Otherwise we would hardly pay to send them into space to use tremendously valuable scientific instruments.

Now we may picture the Space Telescope in orbit, 375 miles high (Figure 6), filling the giant cargo bay of the Space Shuttle, forty-five feet long and fifteen feet across. The Space Telescope was designed to fit as snugly as possible, so there is barely a cubic foot of extra space in the bay. Most Space Shuttle flights carry several satellites and the equipment for several experiments; the manifest for the Space Telescope shuttle flight shows only one cargo, the precious Space Telescope.

As commands are sent to the astronauts to begin to deploy the Space Telescope from the cargo bay, the engineers are thinking about problems that could arise. Will the telescope operate as it was designed to? The engineers worry particularly about the unanticipatedly long time—

Figure 6: An artist's conception of the Space Telescope in orbit, 375 miles above the Earth's surface. *NASA*

close to four years—that the Space Telescope had to wait since its construction was completed in the fall of 1985.

During that time, it had to survive in an environment for which it was never intended—the Earth's surface. Here on Earth, the Space Telescope experienced a host of influences capable of destroying its precision forever, influences that had to be fought—at a cost of millions of dollars per month—as best we could. Of these difficulties, two stand out as particularly devastating: gravity and moisture.

Gravity is an obvious problem. Weighed down by the Earth's gravity, the lightweight trusses that form the internal structure of the Space Telescope were in danger of sagging, not to any visible degree, but enough to destroy the incredibly fine alignment of all the parts of the telescope. Without this alignment, the Space Telescope could never achieve the accurate, crisp views of the universe for which it had been designed. In orbit, the Space Telescope would sail free of the stresses caused by gravity, because all parts of it would be falling around the Earth at precisely the same speed. Even though the Earth's gravity pulls on the telescope (or else it would hardly orbit the Earth), the fact that the Space Telescope orbits in "free-fall"—with no part of the telescope supported any more than another—means that there are no forces on it which might bend it even slightly. But on the Earth's surface, there is no way to support every part of an object equally well; instead, we support it at various points, and the object's internal structure must hold up the rest, like a bridge that spans a body of water. By adding new points of support, we can reduce the distance across which the structure must hold itself up, but never completely. Hence any object resting on the Earth's surface is inevitably subject to gravitational forces that tend to bend it out of shape. So long as the Space Telescope remained on the ground, gravity was bending every part of it, including the finest telescopic mirror ever made.

And what a mirror! The Perkin-Elmer Corporation had put fifty employees to work for three years to solve the problem of creating a mirror on the ground—where gravity tends to bend it—that must function in space, where the free-fall, "weightless" conditions are quite unlike those on Earth. Crafted of special ultra-low-expansion glass, so that it will neither expand nor contract significantly in space, the mirror consists of honeycomb cores, with most of the glass behind the reflecting surface removed in order to lower the total weight of the mirror. Another useful property of the mirror glass, made by the Corning Glass Company, is its weldability, which allows a single slab to be fused from individual honeycombs. As a result of this new technique, the 94-inch-diameter primary mirror weighs only a quarter of what a solid glass mirror would: just under one ton (Figure 7). In comparison, the mirror of the famed 200-inch telescope on Palomar Mountain weighs more than fourteen tons—fourteen times more weight in a mirror that has only a bit more than twice the diameter.

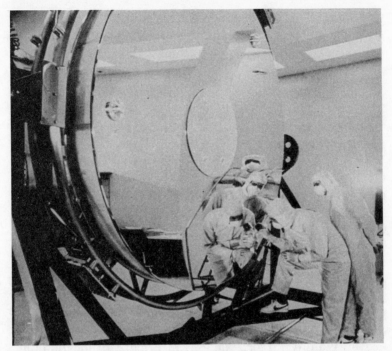

Figure 7: The Space Telescope's ninety-four-inch-diameter mirror.
NASA

The Space Telescope's mirror is strong enough to maintain its shape precisely while in free-fall in an orbit around the Earth, though it would not be strong enough for a telescope on the ground. However, by the time of launch, this mirror had spent several years on the ground, with gravitational stresses upon it. The technicians maintaining the telescope therefore had to try to assure that no permanent sag would arise in the mirror—not even by a millionth of an inch, which would ruin the usefulness of its silver-coated reflecting surface, curved into a near-perfect surface to direct light toward the secondary mirror and on to the telescope's instruments. Any forces that bent the mirror by a millionth of an inch would wreak such damage as to make the launch a useless endeavor, placing into orbit a subpar instrument, hardly an improvement over Earth-based telescopes.

If the technicians succeeded in their fight against gravity, they had to confront an even greater problem: contaminants in the air. Space contains only tiny amounts of hydrogen, oxygen, and nitrogen molecules and atoms. The Space Telescope was not designed to withstand the dust prevalent in the Earth's lower atmosphere, which can settle on the telescope's mirror, thus ruining the mirror's near-perfection.

The solution, as well as it could be achieved, was to keep the telescope in the cleanest environment humans can produce—nowhere so clean as the environment in space, but the best we can do. At the Lockheed Missile and Space Company's facility in Sunnyvale, California, the world's largest clean room housed the Space Telescope during its wait for launch. The Space Telescope's ninety-four-inch primary mirror had received its reflective coating, which turned a disk of glass into a high-efficiency mirror, in December 1981. No one then believed that the mirror would spend more than eight years on the ground before reaching orbit, where it could finally begin to reflect light from cosmic objects, free from airborne contaminants. But so it was, and those who guarded the health of the Space Telescope had to care for the mirror during these eight long years and more, most of it while the telescope remained in the clean room (Figure 8).

If dust did settle on the mirror, the telescope technicians knew that they could attempt to blow it off gently with jets of nitrogen. This they had done before they sent the telescope to Cape Canaveral, but there was no way to be sure that the job was entirely successful until the spacecraft reached orbit, became fully deployed, and began to observe the cosmos. For this among other reasons, the moment of "first light"—the time when the Space Telescope attempts to let light from a celestial object shine on its mirror and reflect to its instruments, with their readings telemetered to Earth—must be the ultimate moment of truth, testing whether or not a decade and more of effort were in vain.

In addition to the problem of dust settling on the mirror,

Figure 8: The Space Telescope in the clean room at Lockheed's facility in Sunnyvale, California. *NASA*

moisture around the Space Telescope played an insidious role. The entire backbone of the telescope—the support structure that might have bent slightly during its years on Earth's surface—consists of a marvelous substance called graphite epoxy. NASA developed this substance during the early 1970s as it sought to find a material that would be lightweight yet immensely strong per pound. Furthermore, NASA sought a substance that would barely expand or contract as it passed through extremes of temperature, from the familiar ones on Earth's surface to a hundred degrees less or a hundred degrees more, as an object in space orbits the Earth and passes first through the Earth's shadow and then out into the bright sunlight. Thanks to these wonderful properties, graphite epoxy is favored for tennis rackets and golf clubs—even for high-priced bicycles—as well as for the framework of a spacecraft.

Graphite epoxy satisfied NASA's exacting specifications brilliantly. But it has a flaw, one that was little recognized until the Space Telescope had to undergo its long wait.

Graphite epoxy absorbs moisture from the atmosphere, incorporating the water-vapor molecules into its rigid molecular structure. Within the clean room, with its relative humidity of about 50 percent, there was plenty of water vapor to pass into the graphite epoxy that forms the Space Telescope's framework. Because launching the Space Telescope into orbit will send it into a realm where essentially no water vapor exists, the water vapor temporarily locked within the graphite epoxy will start to evaporate. This evaporation has a tiny but calculable effect on the structure of the Space Telescope, which will shrink slightly, like a garment on a clothesline.

The evaporation of the water in the epoxy also creates a worse problem. As the water leaves the epoxy, some of it may settle on the surfaces of the telescope, on its mirror, and—even worse—on the fine guidance sensors that detect the positions of guide stars to help point the telescope properly. Small though the amount of water vapor may be, the net effect could be to coat the fine guidance sensors in ice, like an automobile's windshield on a sleety day, leaving them unable to perform their essential function.

Since moisture could ruin the support structure of the Space Telescope, you might expect that Lockheed's clean room would have been maintained at the lowest relative humidity that could be achieved. However, this would have produced an unwanted side effect: static electricity. Once the relative humidity of air falls below 50 percent (that is, once the amount of water vapor in the air is less than half the total that the air can carry before it rains), static electricity occurs readily, as electrical impulses jump between one object and another. At higher levels of relative humidity, the water vapor in the air significantly inhibits such discharges. Since the Space Telescope could not simply be left alone, but had to be tended by a host of technicians to monitor and improve its internal functioning, NASA could not afford to risk any electrical discharge that might ruin the Space Telescope's sensitive electronics. Hence the relative humidity in the clean room stayed above 50 percent—and the risk of moisture being absorbed by the support structure likewise increased.

Quite aware of this problem, and equally worried about it, the Space Telescope's engineers attempted to resolve it. As close to launch as possible, the technicians arranged to place the entire Space Telescope inside an airtight container and to "purge" the system with nitrogen—to drive out all of the air with nitrogen gas under pressure. This would force out not only the water vapor in the air but also the water vapor that had been incorporated into the graphite epoxy. To the extent that the technicians succeed, the Space Telescope can ride into orbit with relatively little water vapor ready to emerge once the telescope reached the near-vacuum of space.

But did they succeed to the point where the fine guidance sensors would work? Only the actual experience can tell.

As soon as the Space Shuttle reaches orbit, the astronauts will begin the procedures that lead to setting the Space Telescope free of the shuttle. This will be the first deployment of such an enormous payload, but not (in plan) the last: future astronauts can visit the Space Telescope to repair and to upgrade it. But first they must make it work. On the day after launch, it will be time to take the instrument from its launch cradle and set it free (Figure 9).

The first step in this process will be to move the entire

Figure 9: An artist's conception of the deployment of the Space Telescope from the Shuttle orbiter. *NASA*

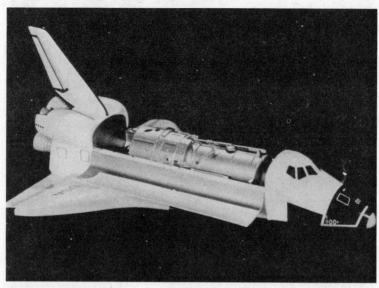

The Space Telescope is in the cargo bay of the Space Shuttle, with the doors open.

The Space Telescope is being deployed from the cargo bay.

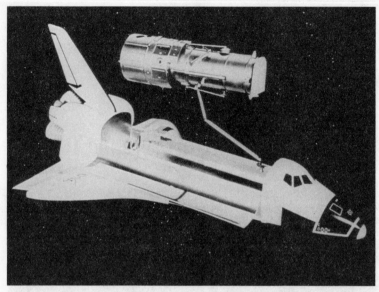

The Canadian-built boom of the Space Shuttle releases the Space Telescope into space, where it orbits independently.

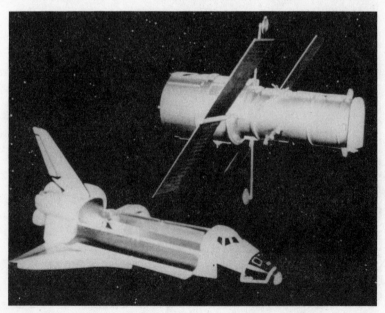

The Space Telescope orbits with its solar panels unfolded.

telescope, as large as a semitrailer, out from the Space Shuttle's cargo bay to a position several hundred yards from the shuttle. Despite being "weightless," the Space Telescope cannot be flung around in space at whim. The *mass* of the telescope does not change by being placed in orbit, and the mass—how much matter any object contains—determines how much it resists being accelerated: despite what you may see in science-fiction films, you can't toss steel beams around in space any more than you can on Earth.

The Space Telescope's mass of nearly eleven metric tons (11,000 kilograms, or 25,500 pounds) makes it a challenge to move about, but NASA engineering has proved equal to this task. The Space Shuttle comes with a Canadian-built boom to move objects out of the cargo bay, slowly and carefully. Although the Space Telescope ranks among the most massive objects the boom has moved, it can do so and will gently maneuver the enormous cylinder to a position outside the cargo bay. Then, with a final slow-motion push, the boom will release the Space Telescope, which will move slowly away, at a speed of about one foot per second.

Once the telescope has moved a few hundred feet from the Space Shuttle, ground commands will start the deployment of the solar panels, which lie curled in their booms, as carefully rolled as can be. The astronauts will hover in the vicinity to help if any problem arises.

The solar panels were designed and built by the European Space Agency to perform the essential task of providing the electrical power that allows the telescope to work. Since the telescope spends half of every orbit—forty-eight minutes out of each ninety-six—in a position where the Earth blocks out direct sunlight, the Space Telescope requires not only solar power but also storage batteries, which provide power through the telescope's forty-eight-minute "night." The batteries draw their power during the forty-eight-minute "day" from the solar panels, silicon wafers that generate electrical current when sunlight strikes them.

The solar panels will unfold along two booms, one on each side of the telescope, that hold the panels in a "double roll-out array"—a fancy way of saying that the panels reach space rolled into tight cylinders, and then unfurl in two opposite directions from each boom. As if in attendance on a giant insect whose new-spread wings are drying, the astronauts will supervise the operations of spring-loaded stiffening braces to help the solar panels become rigid. With the solar panels in place, the Space Telescope will be ready to orient itself properly, with its panels always held more or less perpendicular to the sun as the telescope's mirror points toward an object under observation.

This orientation is more difficult to achieve than it may seem. The Space Telescope continuously circles the Earth, which means that at every moment of its ninety-six-minute orbit, the spacecraft must somehow manage to control the pointing of the telescope itself and of its solar panels in a way that allows the telescope to continue to point—with incredible accuracy—toward a particular object in space while simultaneously keeping the solar panels roughly perpendicular to the sun's rays, at least during the half of its orbit when the telescope is in daylight. These requirements can be fulfilled, thanks to the spacecraft's on-board computer, which is programmed before each orbit to tell the Space Telescope just how to keep turning itself to maintain its observational posture while continuing to draw power through its solar panels.

Every arduous step on the journey toward the success of the Space Telescope—the mirror, the framework, the pointing, and a host of others—required a solution grounded in technological ingenuity and innovation. The Space Telescope broke new ground in optics, in detectors, most of all in pointing accuracy. And yet, as the Space Telescope stood poised for launch early in 1990, the scientists who would ultimately decide whether or not to call for its launch faced a dilemma. On the one hand, the nearly unbearable wait was over. The Space Shuttle was back in operation; the

legion of problems with the telescope itself had been overcome; the public was waiting; the astronomical community
was waiting; the telescope was costing millions of dollars
per month to maintain on the ground. Every instinct, every
scientific consideration called for launch as soon as possible. All but one: the coming of solar maximum.

Solar maximum refers to the maximum of solar activity,
which peaks about every eleven years. Solar activity includes the appearance of sunspots and active regions, as
well as the occurrence of solar flares. All of these phenomena are due to a magnetic field deep in the sun, which
mysteriously emerges every eleven years, reducing the
temperature of the solar surface in localized areas to produce sunspots and releasing energy in the form of x-rays
and ultraviolet light in the corona above. In the active
regions near sunspots, the energy sometimes explodes in
the form of solar flares, which in a few minutes release
large quantities of energy in x-rays and fast particles.

The next maximum in solar activity is due in mid-1990,
about the time the Space Telescope is to be launched.
Already we are building toward the maximum. In March
1989, one of the largest groups of sunspots ever seen
emerged, and a solar flare emitted vast numbers of energetic particles that caused brilliant displays of northern
lights when the particles entered the Earth's atmosphere
(Figure 10).

What does solar activity have to do with the Space
Telescope? The link is the upper atmosphere of the Earth,
through which the Space Telescope will orbit. Although
the Earth's atmosphere at that altitude is very tenuous, the
air resistance, or drag, it produces on the Space Telescope
will, over long periods of time, cause its orbit to decay.
That is, drag will cause the Space Telescope to spiral
slowly but surely toward the Earth, finally to burn up in
the lower atmosphere. Because solar activity directly influences the density of gas in the upper atmosphere, it affects
the amount of drag and hence the orbital lifetime of the
Space Telescope. As an example of this effect, NASA's

Figure 10: The northern lights, or aurora, caused by energetic particles from the sun that strike the atmosphere, guided by the Earth's magnetic field. A radio telescope is in the foreground.

Solar Maximum Mission satellite (which, ironically, is studying solar activity) lost nearly half a mile in altitude in a single day as a result of the heating of the upper atmosphere by the solar flare of March 6, 1989.

Solar ultraviolet radiation is absorbed in the upper atmosphere, heating it to about 3,000 degrees Fahrenheit. (This sounds dangerous for astronauts, but the air is so thin that its high temperature has little effect.) Because the amount of solar ultraviolet is greater at solar maximum, the atmosphere gets hotter and expands. Paradoxically this leads to a greater density of gas at higher altitudes, as denser gas below expands upward. The net effect is that the atmospheric density at the 375-mile altitude of the Space Telescope is about ten times greater at solar maximum than at solar minimum. Correspondingly, the atmospheric drag is ten times larger.

Originally the Space Telescope was designed to orbit at a

higher altitude, where atmospheric drag would have a negligible effect. Unfortunately, the redesign of the Space Shuttle after the *Challenger* accident resulted in decreasing the altitude it can reach with the Space Telescope, so atmospheric drag will be a problem. Under typical solar conditions, the orbit of the Space Telescope should last more than ten years if the altitude is 375 miles. Again unfortunately, we seem to be headed into such an intense solar maximum that atmospheric drag could bring the Space Telescope down in only two or three years. That would be a disaster for the whole project.

Solar activity has two other less important effects, both negative. Energetic particles from the sun directly strike the solar panels that supply power to the Space Telescope, degrading their performance. And because the Space Telescope is not symmetric about its midpoint, atmospheric drag acts unevenly on its surface, tending to send it into rotation and making it hard to point as accurately as is needed to get images as sharp as those of which the Space Telescope is capable.

If the Space Telescope is to be a useful scientific instrument, it will have to do battle with solar activity. NASA has several weapons available in that battle. One is to deploy the Space Telescope at the highest possible altitude. That will be done; 375 miles is the maximum that can be achieved. If the orbit of the Space Telescope begins to decay, another option is to plan a later shuttle mission to rendezvous with it, capture it, and reboost it to a greater altitude. This would be a very challenging mission, and an expensive one, but it probably could be done. The third option would simply be to defer the launch of the Space Telescope until solar activity begins to decline, say, until 1991. This option has its proponents, but the decision was no; the launch was scheduled to proceed.

Too much had been invested in the Space Telescope, too much time spent on the ground at enormous expense, to make it feasible to delay the launch past the time when the telescope was ready to operate and the Space Shuttle ready to launch it. Scientists had to balance all that they knew, all that they could speculate, about the oncoming solar maximum and its effect on the telescope, together with the prospects, which are not bad, for being able to move the Space Telescope into higher orbit as its original orbit decays to a slightly smaller size. The key to probable success was to squeeze just a few extra miles of altitude from the Space Shuttle, miles that could add years to the life of the Space Telescope. So, weighing all these concerns, NASA decided to launch, to deploy in space the world's finest piece of equipment to study the cosmos.

And there we can picture the Space Telescope, 375 miles high, falling continuously around the Earth, directed from the ground at astronomers' commands. Lyman Spitzer dreamed big when he and a host of others conceived of what became the Space Telescope. Their success in creating this mighty instrument does, however, leave a few questions unanswered. In particular, we ought to examine the history that led to the development of this mighty $2 billion instrument and ask: Why do we need the Space Telescope?

2

THE DREAM FULFILLED

THE LAUNCH INTO orbit of the Space Telescope, the first true, large astronomical telescope to escape the Earth's atmosphere, fulfills the hopes and dreams of astronomers and engineers. To make such a dream into reality requires more than vision and commitment. It requires people—people who press on with their dream despite setbacks not even envisioned, let alone planned for, while the dream grows stronger. This chapter tells the story of how a dream became real, not exactly as originally planned, but real all the same; how humans, who dream of going to the stars, did even more: they created instruments that could unlock the mysteries of the stars for them.

The vision of sending spacecraft high above the Earth, eventually into orbit around our planet and far beyond it, might have occurred as early as six centuries ago, when the Chinese developed rockets as weapons. However, only during the last century did serious speculation on the possibility of spaceflight emerge. In Russia, just about a century ago, a schoolteacher named Konstantin Tsiolkovsky wrote books that foresaw the day when humans would travel to other planets in rockets and indeed would colonize other worlds.

Fifty years later, not knowing of Tsiolkovsky's works, an American physics professor at Clark University in Worcester, Massachusetts, a man named Robert Goddard (Figure 11), saw the practical possibilities of using rockets to reach

Figure 11: Robert Goddard (1882–1945), the first American rocket developer. *Goddard Space Flight Center, NASA*

high altitudes. Goddard built a liquid-fueled rocket, the first of its kind. (All previous rockets, like the Roman candles at fireworks displays, had used solid fuel, much easier to handle but relatively inefficient.) In March 1926, Goddard launched his first rocket from his aunt's farm near Auburn, Massachusetts, an event that passed almost entirely unnoticed by the world around him. Goddard continued to improve his rockets and to work out the basic principles of rocket flight until his death in 1945.

In that year, a team of German rocket scientists and engineers made their way, at considerable effort, through war-torn Germany, in order to surrender themselves and their stores of information to the advancing American, rather than to the Russian, armies. These German workers had designed and fabricated both the V-1 (essentially a drone airplane, which required air in order to ignite its fuel) and the V-2 (a true rocket, which could soar above almost all of the atmosphere, since it carried its own oxygen for combustion). From their launch sites in Holland and northern Germany, the V-1 and V-2 had rained destruction on targets in southern England—and, owing to their primitive guidance systems, on nearby locales as well. Hitler had hoped that his "secret weapon" would bring England to her knees, since the Allies possessed no such weapons of destruction.

Hitler was wrong, but his team of scientists became the nucleus of the American postwar rocket program, and the V-2 rockets removed from Nazi Germany (almost from under the noses of the Soviet armies who were scheduled to occupy the site of the V-2 production a few weeks later) were delivered to White Sands, New Mexico, where they served as the launch vehicles for the first American rocket flights. Years later, Wernher von Braun (Figure 12), one of

Figure 12: Wernher von Braun (1912–1977), who helped develop German and U.S. rockets. *Marshall Space Flight Center, NASA*

the original German rocket team, recounted how the German scientists were surprised to be interrogated about the source of their basic insights into rocketing. "We learned it all from your Dr. Goddard," they replied, as indeed to some degree they had.

Soon after the end of World War II, the U.S. government created a "rocket panel" of scientists to advise it on how best to use the captured V-2 rockets. Following this panel's advice, the United States embarked on a systematic research program that used the V-2s and added to them an upper stage, the WAC Corporal rocket. In addition, during the mid-1950s, as part of the plan to improve American rocketry, the United States built its own rocket, the Viking.

Like the V-2 with a WAC Corporal as an upper stage, the Viking was eventually used as a multistage rocket, with a large booster for high initial velocity and to carry it to several dozen miles of altitude, where a smaller Aerobee rocket could carry a small payload even higher. The V-2/WAC Corporal and the Viking/Aerobee reached altitudes above 200 miles—high enough to leave most of the atmosphere below. These rocket flights allowed scientists to capture the first photographs of x-rays emitted by the sun, photographs that are impossible to achieve near the Earth's surface, where our atmosphere blocks all incoming x-rays. Since the rocket payloads spent only a few minutes at high altitudes, they could not photograph faint objects such as stars and galaxies in x-rays or in any other type of electromagnetic radiation.

But the x-ray observations of the sun had their own importance. Solar x-rays affect the upper atmosphere by ionizing some of the atoms there—that is, by stripping away one or more of the electrons that orbit the atoms' nuclei. This ionization produces the region in the upper atmosphere called the ionosphere. The behavior of the ionosphere in turn affects how well the Earth's upper atmosphere reflects radio waves. Reflection by the ionosphere bounces relatively long radio waves (including those used for AM radio) to far greater distances from the

radio transmitter than they could reach without iono-
spheric reflection. World War II had shown scientists how
important this effect could be, and any investigation of
changes in the reflectivity of the atmosphere had a claim on
military funding. Hence the Naval Research Laboratory in
Washington, D.C. and its associated institutions grew in
size and importance as rocket investigation of the ionosphere moved forward.

The scientists and engineers who worked to send rockets
above the atmosphere in order to make observations of the
sun knew that many further steps in rocket exploration lay
ahead, including launching a spacecraft into Earth orbit.
Working in collaboration with others, Herbert Friedman,
the head of the x-ray observing program at the Naval
Research Laboratory, created several possible paths that
could lead to the scientists' goal of artificial satellites in
orbit around the Earth. These satellites would be able to
observe the sun and other objects continuously in types of
radiation that cannot penetrate our atmosphere. U. S. scientists planned to launch a twenty-five-pound sphere, to be
called MOUSE (Minimum Orbiting Unmanned Satellite of
Earth) during the International Geophysical Year of 1957-
1958.

Though the scientists' plans were reasonable and defen-
sible, they were rendered obsolete when the Soviet Union
launched the first two artificial satellites, Sputnik I and
Sputnik II, during the fall of 1957. Sputnik I, launched on
October 4, 1957, weighed a then-astonishing 84 kilograms
(185 pounds). Sputnik II, sent into space with a dog, Laika,
on November 3 of that year, had the astounding mass of
1,121 pounds.

History may well record that the Sputnik launches
marked the end of American belief in the innate superiority
of the United States—a belief that has not, of course,
completely disappeared, but that no longer dominates
American thought as it did during the generation between
World War II and 1957. The immediate American reaction
to Sputnik was astonishment: how could the finest of all

countries have missed the boat in launching the first artificial satellite? The next question was even more emotionally
freighted: how soon could the United States place its *own*
satellite in orbit and thereby demonstrate superiority to, or
at least parity with, the Soviets?

No one who lived through the years 1957 and 1958 in the
United States can forget the eagerness and despair caused
by the first attempts to match the Russians by orbiting an
artificial satellite. The initial hopes lay with the *Vanguard,*
a nonmilitary rocket that appeared to be hopelessly too
small, whose construction was directed by the Naval Research Laboratory. On December 6, 1957, an enormous
television audience watched the black and silver *Vanguard*
rise four feet above its launch pad and then explode, dashing hopes for an American satellite in 1957. Finally, the
nation turned to Wernher von Braun and his team of rocket
scientists.

Along with more than a hundred members of the German V-2 rocket team, von Braun had moved to Huntsville,
Alabama, in 1950 to develop rockets for the U.S. Army. By
1957, von Braun's team had developed a single-stage
rocket, named the Redstone in honor of the Redstone Arsenal (part of which became the nucleus of NASA's Marshall
Space Flight Center) where they worked. On October 4,
1957, when the news of the *Sputnik I* launch reached the
United States, Defense Secretary Neil McElroy was in-
specting the Redstone Arsenal, together with the Secretary
of the Army and several important generals. According to
Major General John Medaris, who was then the com-
mander of the Redstone Arsenal, "Von Braun started to talk
as if he had been vaccinated with a Victrola needle . . .
'Vanguard will never make it. We have the hardware on the
shelf. For God's sake, turn us loose and let us do something.
We can put up a satellite in sixty days, Mr. McElroy!' "

More cautious than von Braun, Medaris suggested
ninety days. Five days after the launch of *Sputnik II*, the
order to proceed came from Washington. At the end of
January 1958, less than ninety days afterward, the Huntsville team used a modified Redstone rocket (the Jupiter-C)

to place a small American satellite, *Explorer I*, into orbit around the Earth.

Explorer I had only a few instruments on board, but it was a true Earth-orbiting satellite, and it belonged to the United States. The nation was back in the space race, and in a big way, even though the satellite itself was small. One of the instruments on *Explorer I*, a device designed by Professor James Van Allen of the University of Iowa (Fig-

Figure 13: James Van Allen (1914–   ), a pioneer in exploring space near the Earth.

ure 13) to detect energetic particles, was nearly jammed by huge numbers of particle impacts as it passed through what was previously believed to be empty space above the Earth's atmosphere. We now know that the energetic particles detected by Van Allen's instrument are trapped by the Earth's magnetic field in what physicists call the Van Allen radiation belts (Figure 14). The United States' first scientific payload in space had made an important discovery, one that evolved into a new field of science, now called the physics of the Earth's magnetosphere. But the scientific advances from *Explorer I* continued to take a back seat to the "big story," the race between the Soviet Union and the United States for "superiority" in space.

In the meantime, the U.S. government had acted decisively to provide for the longer-term aspects of space exploration. Out of an advisory board called NACA, the National Advisory Committee on Aeronautics, the Eisenhower administration and the Congress shaped NASA, the

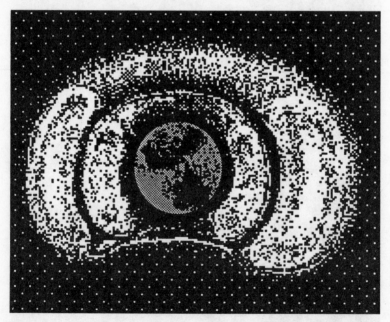

Figure 14: An artist's representation of the Van Allen radiation belts, which contain energetic particles trapped in the Earth's magnetic field.
Drawing by Marjorie Baird Garlin

National Aeronautics and Space Administration. NASA came into existence in 1958 with a charter to explore space and the funding to do so seriously. In the same year, the National Academy of Sciences (a body chartered in 1863 by Congress to further scientific knowledge and to advise the government on scientific research) created a committee of scientists, the Space Science Board (now called the Space Studies Board), to encourage the participation of scientists in the emerging space program. The board immediately gathered proposals suggesting scientific investigations that could best be performed with spacecraft. Thus in October 1958, when NASA began its formal existence, serious proposals for scientific work in space had already been prepared.

The astronomers proposed a program to observe the cosmos in ultraviolet light, using automated, unmanned telescopes in relatively low Earth orbit, a few hundred miles up. These telescopes would be stabilized in their orientation, so that they could point toward any object in the sky and could observe for long enough times to gather significant information. Other specialized instruments would observe the sun with equipment quite different from that used to observe the stars, since a solar telescope receives millions of times more light than the stars send us, and it must focus on extended areas of the sun's surface, rather than on stellar points of light. Because the requirements for solar astronomy appeared to differ so greatly from those for stellar astronomy, the astronomers proposed that Orbiting Solar Observatories (OSOs) should be developed separately from Orbiting Astronomical Observatories (OAOs). Following these recommendations, NASA eventually developed a long and successful series of Orbiting Solar Observatories that examined the sun's ultraviolet light in detail (Figure 15).

SOLAR ASTRONOMY FROM SPACE

The OSOs confirmed that the visible sun lies within a tenuous envelope (called the solar corona) that has a

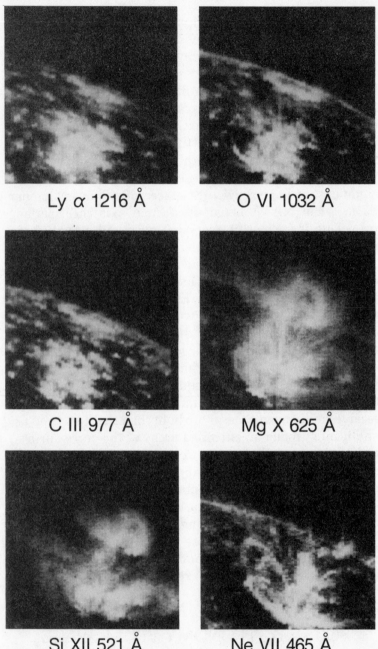

Ly α 1216 Å   O VI 1032 Å

C III 977 Å   Mg X 625 Å

Si XII 521 Å   Ne VII 465 Å

ACTIVE REGION, EAST LIMB
NOV 23, 1973 0552-0736 UT

Figure 15 (opposite): Ultraviolet images of an active region on the sun. Each image was taken at the wavelength of light emitted by a different chemical element, each of which emphasizes a different range of gas temperatures. Coronal loops are evident in the image taken in the light of six-times-ionized neon, called neon 7 (abbreviated here as Ne VII). *Harvard College Observatory*

temperature of nearly four million degrees Fahrenheit—far higher than the ten thousand degrees of the solar surface, or photosphere, which is the part of the sun that we would see if we were to look at the sun with the unaided eye (*not a recommended activity*, since the sun will damage the eye's retina in less than a second!).

Although astronomers had observed the solar corona long before the OSO satellites, the ultraviolet observations made from the OSOs demonstrated that the hot gas in the corona mainly concentrates in "coronal loops," clouds of gas concentrated where solar magnetic fields arch into the corona from the underlying photosphere (Figure 16). Coronal loops often connect two or more of the dark regions on the solar surface called sunspots.

Where does the sun's magnetic field come from, and why should the coronal gas be concentrated by the magnetic field? Ultraviolet observations of coronal loops enabled astronomers to identify within them a variety of different chemical elements, atoms ionized by collisions with fast-moving particles in the hot coronal gas. For example, oxygen ions that retain only three of oxygen's ordinary complement of eight electrons appear in great numbers in the coronal loops. By analyzing the numbers of the various ions, solar astronomers determined both the temperature of the coronal gas and its density, or the number of atoms in each cubic inch. From the fact that the density and temperature vary strongly along each solar loop, astronomers conclude that heat typically flows from the four-million-degree gas within the loops down into the much cooler photosphere, much as the heat from your hand flows into a piece of ice that you hold.

Since heat is a form of energy, observations of heat flow

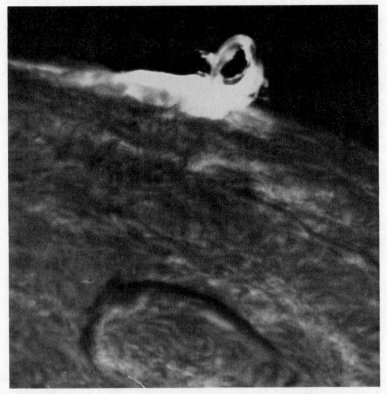

Figure 16: The sun's surface, photographed in the light of hydrogen gas. A bright coronal loop of gas is trapped in a magnetic field above the surface. *Big Bear Solar Observatory*

raise the question, Where does the energy of the hot coronal gas come from? Since energy can't simply be created from nothing, the energy in the corona must have a source—but none is visible in the corona itself. Astronomers now believe that the energy required to explain the flow of heat out of the corona is stored invisibly in electrical currents that create the magnetic field observed in the solar corona. By some mechanism not yet understood, the energy of these currents can heat the gas in the corona, just as the electrical currents that dissipate within the fine wires of a toaster heat them and the volume inside the toaster. The sun's corona resembles a giant toaster, emitting energy at the rate of a billion trillion watts. Needless to say, solar astronomers remain extremely interested in how and why this occurs.

While the ultraviolet astronomers were observing the sun during the 1950s and 1960s, the x-ray astronomers were perfecting methods to obtain x-ray images of the sun. This is not an easy task: x-rays pass straight through most objects, rather than reflecting from them, as one desires in order to focus x-rays with the mirror in an "x-ray telescope." Since x-rays that pass through objects are barely deflected, astronomers likewise find it difficult to construct an "x-ray lens." The problem of focusing x-rays was solved by placing the mirror of the x-ray telescope nearly *parallel* to the incoming beam of x-rays. If x-rays strike such a mirror at a sufficiently shallow angle, rather than almost perpendicularly, they will indeed reflect from the mirror. Therefore a "grazing-incidence" mirror—a mirror that reflects at only a small angle—lies at the heart of every x-ray telescope.

By 1973, when NASA's *Skylab* satellite was sent into orbit, using an enormous Saturn 5 booster originally developed for the Apollo project, ultraviolet and x-ray telescopes were both ready for the astronauts to carry up with them. *Skylab* had problems from its start: One of the panels designed to fold out in orbit to collect solar power to run the spacecraft ripped off during the launch. The other solar panel was damaged and did not deploy properly. During the two weeks before the astronauts themselves were launched for rendezvous with *Skylab*, NASA engineers worked furiously to design a way to fix the damaged solar panel. Through their efforts, the astronauts were able to deploy the panel well enough for the mission to succeed.

Skylab carried an Apollo Telescope Mount, a mounting system that compensated for the spacecraft's motion to keep telescopes pointed accurately at the sun for long periods of time. A number of instruments were available to observe the sun's visible light, ultraviolet light, and x-rays. All of them operated successfully, and our understanding of the solar corona improved greatly as a result of *Skylab*'s flight.

One result obtained with the x-ray telescope on *Skylab* remains particularly fascinating: the distinction between coronal loops and coronal holes. X-ray images of the sun (Figure 17) show clearly that although the solar photosphere emits no x-rays (since its temperature is too low to produce them), the corona glows with x-rays because of its high temperature—millions of degrees Fahrenheit instead of the mere thousands of degrees in the photosphere. Furthermore, as had been suspected from earlier ultraviolet and x-ray observations, the sun's x-ray emission arises from coronal loops that align with magnetic fields arching up from the photosphere. The giant loops are evident in Figure 17, and we can even discern smaller loops. The *Skylab* observations confirmed that the structure of the

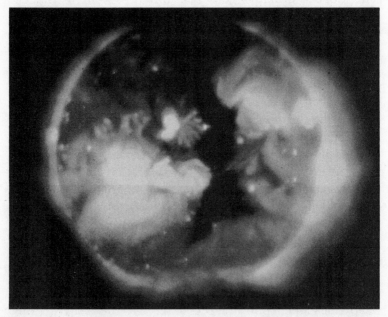

Figure 17: The dark areas on this x-ray image of the sun are coronal holes, regions in the corona of particularly low density. The bright areas are emission by million-degree gas trapped in coronal loops, where the density is high. *American Science and Engineering, Inc.*

sun's corona is highly irregular: the gas is hottest and densest in the coronal loops, where the magnetic field is strong, and cooler and more rarefied where the magnetic field is weaker.

Figure 17 also reveals dark regions, called coronal holes, where any gas that may exist is not hot enough to emit x-rays. Through such coronal holes come the streams of particles that form the solar wind, or at least the fastest-moving streams of the solar-wind particles. Early in the space age, scientists found that the Van Allen radiation belts obtain their energy largely from a solar wind of particles from the sun that blows past the Earth at a speed of about 300 miles per second. This impressive speed arises from the four-million-degree temperature in the solar corona, which gives particles enormous energies as they move in random directions. At such high temperatures, some of the atoms are moving so fast that they can escape from the sun in spite of its gravitational pull.

Magnetic fields pervade coronal holes as well as coronal loops, but instead of reentering the sun as the magnetic fields in the coronal loops do, the fields in the holes wander off into the solar system. Since the ions in the solar corona are electrically charged, they can move only along the local magnetic field, which "enslaves" electrically charged particles with electromagnetic forces, making them follow the field wherever it leads. Because the magnetic field reenters the photosphere in the coronal loops, ions there cannot escape from the sun. But ions *can* escape along the magnetic fields in the coronal holes, which wander out into the solar system. Thus the bulk of the solar wind consists of ions and electrons moving outward along "open" magnetic fields located within the coronal holes.

Astrophysicists have conjectured that electrical currents present in coronal holes dissipate energy there, just as they do in coronal loops. However, in the coronal *holes*, the energy is used up in helping the hot gas to escape from the sun's gravitational pull, whereas in coronal *loops*, the electrical energy is converted into heat. In short, gas es-

capes through the holes but heats up in the loops. Because
the ions in the magnetic field of a loop are trapped there,
their numbers build up to the point that we can see the
resulting x-rays that the hot gas produces.

Our new perception of the solar corona is a fundamental
scientific advance achieved by NASA's program of space-
borne ultraviolet and x-ray astronomy observing plat-
forms. This division into loops and holes has challenged
theoretical physicists to explain how electrical currents can
dissipate both into heat (in the coronal loops) and into
energy of motion (in the coronal holes). This challenge in
turn has led to significant advances in the field of plasma
physics, which studies the behavior of electrically charged
particles in magnetic fields.

Our understanding of solar physics has also allowed
improved predictions of the sun's effects on the Earth's
atmosphere. If the ultraviolet light and x-rays from the
sun's corona did not exist, the Earth would have no ionosphere, so there would be no possibility for AM radio
waves to cover long distances by reflecting again and again
between the ionosphere and the Earth's surface. In addition, the ultraviolet light emerging from the sun's lower
corona (called the sun's chromosphere) produces the ozone
in our atmosphere that protects us from ultraviolet light.
On the other hand, if the corona did not exist, there would
be no solar ultraviolet radiation and hence no "need" for
ozone to protect humans from the ultraviolet component of
sunlight. But since the sun does have a corona and a
chromosphere, we need our ozone layer. Therefore a sufficiently good understanding of the sun's chromosphere and
corona to predict their behavior has proved important for
everyday life on Earth—even more so as we find that
certain synthetic products are destroying the ozone layer.

Although most astronomers failed to recognize it at the
time, NASA's solar observatories were laying the ground-
work for new fields of astronomy: ultraviolet and x-ray
astronomy. For the sun is an ordinary star, of which literally billions exist in our galaxy. If the sun has a corona, no

doubt other stars do as well. Therefore, x-ray and ultraviolet observations should eventually reveal the hot coronas around other stars. Recall that we are speaking here of stars whose surfaces are relatively cool, "only" ten thousand degrees, say, but that have million-degree coronas as a result of their magnetic fields. Farsighted astronomers realized that the study of magnetic fields in other stars could prove extremely useful in understanding the magnetic fields on the sun.

STELLAR ASTRONOMY FROM SPACE

This remark might seem puzzling, for shouldn't we always expect to learn more about the sun by studying it directly, instead of by observing other stars? No, and for the following reason. Magnetic fields of stars are intrinsically highly chaotic, complex, and mixed-up, rather like the weather on Earth. To understand what magnetic fields are like, astronomers need many *different* examples of the phenomenon to study, so that they can derive the average properties of the fields and compare these average values with those predicted by theoretical models. The theoretical predictions depend on certain properties of the star in question—for example, its mass, radius, and rate of rotation. If astronomers study only one star, the sun, they cannot tell whether magnetic fields correlate as predicted with these properties. But with data from a range of different types of stars, astronomers can check their theories to determine which is correct.

Recognizing this fact, astronomers pressed NASA for x-ray and ultraviolet telescopes in Earth orbit to study the stars. However, because other stars are so much more distant than the sun and therefore relatively faint, observing them requires a completely different set of instruments. An identical need for satellite observations of faint objects was stressed by astronomers interested in observing stars so hot that even though they are not magnetic, they emit most of their radiation in the ultraviolet. NASA responded

to these requirements by funding a set of Orbiting Astronomical Observatories (OAOs). One of the results of this
program, as had been hoped earlier by solar astronomers,
has been the creation of a data base on magnetic stars that
already makes it possible to rule out the idea that the
corona is heated by shock waves from the photosphere
rather than by electrical currents.

The stellar-oriented astronomers began work on the
Orbiting Astronomical Observatories as soon as NASA
was established in 1958. They realized that they could
learn a great deal about objects' ultraviolet spectra—the
distribution of energy among different colors of ultraviolet
light—by using telescopes with modest mirror diameters,
even less than thirty-six inches, which represents a small
telescope for an Earth-based observatory. The first successful Orbiting Astronomical Observatory, OAO-2,
launched on December 7, 1968, carried one sixteen-inch
telescope and four eight-inch telescopes. The second, OAO-
3 (Copernicus), launched on August 21, 1972, employed a
thirty-two-inch telescope.

OAO-2 came first and made important advances in ultraviolet observations, but it had nowhere near the observing capability of the Copernicus satellite. The key difference between the two satellites lay in their ability to
perceive the colors of ultraviolet light. Our eyes see different colors of visible light, perhaps several hundred different shades; finer instruments can perceive thousands, even
millions of different colors. These colors have tremendous
scientific importance, because each type of matter in the
universe—each kind of atom, ion, or molecule—affects only
certain colors of light, leaving the others alone. Hence by
analyzing the colors of light from stars and galaxies, astronomers can determine what types of matter exist in far-
distant objects, and even how much of each type of matter
an object contains.

What goes for the colors of visible light also holds true
for the colors of ultraviolet light. Of course, since our eyes
cannot detect ultraviolet light, to us all ultraviolet has the

same color: invisible. But to an instrument capable of analyzing ultraviolet light, each color stands out as distinctly as the shades of red, blue, and green in visible light do to us. And these ultraviolet colors are, if anything, even more densely crowded with information than the colors of visible light. Many types of atoms, ions, and molecules do not interact at all with visible light, but reveal themselves only in ultraviolet. Only in ultraviolet can we see the changes that these atoms, ions, and molecules produce in the light from a star or a galaxy. In short, if you want to find out the full truth about what exists in the universe, you must observe it in ultraviolet, perhaps even before you observe it in visible light.

This is just what the *Copernicus* satellite did. For five years, the satellite observed star after star, region after region in space, storing its data briefly and then sending it back to Earth for analysis. The ultraviolet colors revealed by *Copernicus* painted a new picture of the universe, allowing us to detect previously unseen types of atoms, ions, and molecules in space.

From the success of the *Copernicus* satellite grew the plans to create the Space Telescope, a single instrument with two great abilities: It would have all the capability (and far more!) of any previous ultraviolet-detecting satellite, with a larger telescope and more sensitive detectors than those on *Copernicus*. Also, it would be the first large visible-light telescope in orbit high above the atmosphere, where the telescope could profit from the much sharper views available in the absence of atmospheric blurring and distortion. This was the great space project of the late 1970s, a project that, however, stretched all through the 1980s.

The most used ultraviolet-observing satellite (before the Space Telescope), the *International Ultraviolet Explorer (IUE)*, was launched on January 26, 1978, and carried an eighteen-inch telescope (Figure 18). A dozen years later, the *IUE* is still operating high above the Earth. Moving in synchronous orbit 22,000 miles above the Earth, the *IUE*

Figure 18: The *International Ultraviolet Explorer* (IUE) satellite being prepared for launch. *NASA*

remains poised above a particular point on the equator, sending data back to its creators, NASA and the European Space Agency.

Because of its synchronous orbit, the *IUE* can be operated more conveniently than the Space Telescope can. The *IUE* is visible at all times from the Goddard Space Flight Center, so astronomers there can operate it in real time, sending commands up to the *IUE* and getting data back immediately. This will not always be possible with the Space Telescope, because during a certain fraction of each orbit, the telescope will be out of contact with the ground.

IUE observers have found the satellite a joy to use. With the *IUE*, spectra of the objects under observation are continuously displayed on a television monitor at the Goddard Center so that observers can modify their observing plans in real time in response to what they see. For example, if an object proves to be brighter at the ultraviolet wavelengths at which an observer is studying it than was expected on the basis of previous data, the observer may finish the observation in a shorter time than planned, having accumulated enough information, and can then move on to the next target. This can be accomplished by sending signals to the spacecraft directly from an antenna at the Goddard Space Flight Center.

An astronomer using the *IUE* satellite therefore resembles a teenager playing a video game, but in the case of *IUE*, the player operates a real spacecraft, twenty-two thousand miles high, that moves when he or she presses the button. Science has many advantages; one of them is that sometimes an astronomer can get an easy thrill even while studying the cosmos.

THE GENESIS OF THE SPACE TELESCOPE

The scientific concept of a telescope in space first saw light in 1946, when Lyman Spitzer proposed a Space Telescope with a mirror somewhere between sixteen and fifty feet in diameter. Throughout the 1960s and 1970s, as the ultraviolet observatories *OAO-2*, *OAO-3*, and *IUE* were being planned and built, the Space Telescope remained for astronomers more a dream than a realizable project. The difficulty lay in a fact readily apparent to all: To construct a telescope much larger than the OAOs or the *IUE* would cost far more than those satellites. Just the same, astronomers needed a large telescope to extend their ultraviolet studies to fainter objects than could be studied with smaller telescopes such as the OAOs and *IUE*.

During the 1960s and 1970s, Lyman Spitzer continually reminded the astronomical community that even a ten-foot (120-inch) telescope in orbit could obtain far sharper images than those available from the much smaller telescopes on the OAO and *IUE* satellites. Spitzer also emphasized to his colleagues that a telescope in orbit can observe much fainter objects than any ground-based telescope. This is true even though eleven ground-based telescopes have mirrors larger than a diameter of ninety-four inches (the mirror size finally chosen for the Space Telescope), because the Space Telescope's sharper images stand out much better against the sea of background light and allow the detection of fainter objects. Therefore, Spitzer argued, a Large Space Telescope could provide better performance than any Earth-based telescopes in three different respects: ultraviolet observation, limiting object faintness, and sharpness of images. However, the leading astronomers who used ground-based telescopes were not particularly enthusiastic, largely because they had their hands full with a host of challenging projects for their Earth-based instruments that they could finish without the Space Telescope.

But as the years passed, astronomers came to realize that a Space Telescope would provide an immensely powerful instrument for the future. In 1962 and again in 1965, the National Academy of Sciences convened studies to discuss priorities in space science, both of which studies examined the proposed Space Telescope. Spitzer himself chaired a third National Academy of Sciences study, whose report was published in 1969, this time devoted entirely to the Space Telescope. Because the participants in the 1969 study consisted mostly of the leading users of ground-based telescopes, the fact that this group also endorsed the concept of the Space Telescope proved particularly significant.

The next key step toward the Space Telescope was to

convince the entire astronomical community that the project had such great importance that it deserved priority over other projects. This issue arose when the National Academy of Sciences convened an Astronomy Survey Committee in 1970 under the chairmanship of Jesse Greenstein of the California Institute of Technology, and charged it with the task of setting priorities for all of American astronomy in the 1970s. Of the twenty-three members of the Greenstein committee, only six, including Herbert Friedman, Leo Goldberg, and Lyman Spitzer, had a substantial commitment to space astronomy. Nevertheless, when the Greenstein committee issued its recommendations in 1972, a 120-inch-diameter Large Space Telescope emerged on the list of seven items of "high scientific importance."

To be sure, the committee report stated that funding for these seven, "although urgent, should not create a delay in funding" for four items of still higher priority. These top four items included the Very Large Array of radio telescopes, a program in ground-based optical astronomy, an increase in support for infrared astronomy, and NASA's series of High Energy Astronomical Observatories (HEAOs) to study x-rays and gamma rays from space. All four bore significant fruit during the 1970s and 1980s.

The relatively low priority given the Space Telescope in the Greenstein report reflected the fact that the *OAO-2* satellite was the only one of NASA's ultraviolet space telescopes that had been launched when the report appeared in 1972. Hence scientists were concerned that an immediate leap to a 120-inch-diameter telescope would be feasible only if substantially greater funding became available for astronomical research. The Greenstein report suggested that if the Space Telescope were to be launched during the early 1980s, NASA's annual budgets for astronomical research would have to climb from $20 million in 1970 to $200 million in 1980. The report recommended that as an intermediate step toward the 120-inch Space Telescope, a prototype telescope about sixty inches in diameter be constructed and launched to test and prove the concept of the larger Space Telescope. Later, individual members of the Greenstein committee gave strong support to the Space Telescope in a letter made available to Congress.

During the 1970s, proceeding along the lines delineated by the Greenstein committee, NASA launched the OAOs and conducted other ultraviolet astronomy efforts; it also funded several contractors' studies of how to fabricate telescopes capable of producing the sharpest possible images. Among the contractors was the Perkin-Elmer Corporation of Danbury, Connecticut, which was rumored to be building optical spy satellite systems for the Department of Defense and the CIA. For spying on foreign military operations (if any such activities were indeed occurring), an intelligence-gathering network requires optical systems able to derive the maximum information from each image. Such optical systems, in scientific jargon, would produce "more bits for the buck." Thus more than coincidence would be involved if NASA were to call upon contractors who had built spy satellites to investigate the design of equally capable telescopes that would be directed to look outward at the universe rather than downward at the Earth.

During this decade of development (1972–1981), ultraviolet astronomy at NASA was the bailiwick of Nancy Roman, a stellar astronomer who had made important discoveries concerning the abundances of the chemical elements in different types of stars. Dedicating herself to improving astronomical facilities in space, Roman spent seemingly endless hours behind her desk at NASA and in sometimes stultifying committee meetings to seek the proper scientific goals, the technological means, and the budgetary strategems that would bring the Space Telescope into being.

While serving on a NASA advisory committee early in the 1970s, one of us (George Field) was surprised to hear Roman say that NASA was considering a rapid and direct move toward the initiation of the Space Telescope project in

the mid-1970s, rather than accepting the less expensive
option offered by the Greenstein committee. NASA was
opting instead for a much larger budget, in line with the
most optimistic projections in the Greenstein committee,
according to which expenditures would increase by nearly
ten times to construct the Space Telescope.

A major reason for this action was that NASA studies
had shown that much of the expense of space-borne telescopes goes into subsystems such as power, communications, and scientific instruments that would be the same
whether the telescope itself was large or small. Hence it
made sense to build the largest telescope the budget would
allow. In any case, the astronomers were delighted.

Early in the 1970s, NASA swung into action, and a plan
was soon in place. A "phase A study," meant to demonstrate the feasibility and desirability of the Space Telescope, was completed within a year. A "phase B study,"
aimed at creating a detailed engineering design, was begun
in 1973, and one of us (George Field) became a member of
NASA's Space Telescope Working Group, formed to advise
on the design study. This group was chaired by C. R. "Bob"
O'Dell, the former director of the Yerkes Observatory of the
University of Chicago, one of the unsung heroes of the
Space Telescope. As NASA's Chief Project Scientist for this
most complex of projects, O'Dell was responsible for assuring that the telescope met all the scientific requirements.
Working with scientists experienced in space astronomy,
with technical experts whose advice would be helpful to
the project, and with astronomers who were expected to
have a broad view of the directions in which astronomical
research was and should be heading, O'Dell supervised the
Space Telescope Working Group as it met frequently between 1973 and 1977 to discuss the project in detail, striv-
ing to make the Space Telescope the best instrument possible within the budgetary and design constraints.

One key decision for the working group concerned the
size of the telescope mirror, originally envisioned as 120
inches in diameter. Although the Space Shuttle might have

accommodated such a mirror, the working group learned that substantial dollar savings (and a consequently increased likelihood of funding for the project) would arise from scaling the mirror back to ninety-four inches. The astronomers reluctantly agreed to the smaller size, and ultimately the word *large* was dropped from the Space Telescope's name. This scaling down caused the Space Telescope to emerge with a collecting area only about two-thirds as large as that originally planned. The reduction in size meant that the Space Telescope would take one and a half times as long as originally planned to collect the same amount of light from any given object.

Everyone knew that the Space Telescope, whether or not prefixed by the word *large*, would be expensive. Experience with ground-based telescopes had shown that doubling the diameter of a telescope's mirror would multiply its cost not by a factor of two, but by nearly a factor of eight, the cube of two. It could then be argued that going from the thirty-two-inch diameter of the *OAO-3* telescope to the ninety-four-inch diameter of the Space Telescope, nearly three times larger, would increase the cost by a factor of three cubed, or twenty-seven. Since the cost of *OAO-3* was about $240 million (in 1986 dollars), this adjustment factor suggested that the Space Telescope would cost twenty-seven times $240 million, or about $6 billion in 1986 dollars. The actual amount spent to construct the Hubble Space Telescope reached almost $2 billion in 1986 dollars. This amount falls far below what might have been estimated on the basis of *OAO-3*. However, $2 billion still represents a considerable sum of money, arguably the most ever spent for a scientific instrument. (The Superconducting Supercollider, to be built around Waxahachie, Texas, during the next decade, will shatter this record when it is completed.)

The $2 billion spent on the Space Telescope represents a capital expenditure, analogous to the cost of building a house. But like a house, or like a ground-based telescope, the Space Telescope is built to last—with good luck—for fifteen years or more. Over those years, the telescope will perform many thousands of observations, and while it is operating, the salaries of perhaps as many as a thousand workers involved in its operation must be paid. These people include the scientists, engineers, telescope operators, and computer specialists at the Goddard Space Flight Center who must monitor the safety of the spacecraft, enter the commands that guide it, and plan future changes of the instruments on board. They include the astronomers at the Space Telescope Science Institute who must keep detailed watch over the performance of every aspect of each scientific instrument on the Space Telescope, in order that astronomers elsewhere who use the Space Telescope can draw the most from their observations. And it includes the programmers who will constantly be improving the software for retrieving and processing the data obtained with the Space Telescope.

The total amount to be spent on the Space Telescope also includes the costs to astronomers in universities and elsewhere who will use it, costs such as the expense of computers to process images, of graduate students to help process the data while being trained in the best way to use the Space Telescope, of travel expenses to visit the Space Telescope Science Institute, and of the charges made by scientific journals that publish the observations. Finally, the total cost includes the outlays for developing entirely new instruments to be installed in the Space Telescope, for replacing existing instruments that have worn out or become obsolete, for training astronauts to rendezvous with the Space Telescope to repair it and to replace the instruments, and for sending the astronauts into orbit in the Space Shuttle to do just that.

NASA now estimates that the Space Telescope's yearly operating costs will total $150 million. This figure may be compared with the $22 million annual operating cost of the National Optical Astronomy Observatories, which include two ground-based 160-inch-diameter reflecting telescopes. One thing is certain: the Space Telescope is big science.

WINNING APPROVAL FOR THE TELESCOPE

The Space Telescope is one of the many projects in NASA's budget, which currently totals over $10 billion per year. This sum, large though it is, amounts to 1 percent of the total federal budget and about 35 percent of that 1 percent is funded through deficit spending. Thus, about a third of NASA's budget is, in effect, financed by borrowing. As this borrowing is done to a large extent overseas, especially in Japan and Europe, the Space Telescope depends heavily on the U.S. government's willingness and ability to continue to attract foreign loans. Part of the process of gaining approval for the Space Telescope involved inviting the participation of the European Space Agency, which directly contributed 15 percent of the total cost of the project. But because of the federal budget deficit, the project is in a sense an international one anyway, because foreign loans have helped to finance it.

Since NASA forms an integral part of the U.S. government, NASA's budget must be proposed by the president and approved by Congress. For the NASA budget in general and the Space Telescope project in particular, several officials played key roles in this funding process. During the 1970s, NASA's Administrator, James Fletcher, had to be convinced of the Space Telescope's merit by NASA's Associate Administrator for Space Science and Applications, Noel Hinners. Hinners in turn responded to proposals from the Astrophysics Division, headed by Bland Norris, and from Nancy Roman, who was then in charge of the ultraviolet astronomy branch.

In 1973, with the help of his staff and in consultation with the Space Science Board at the National Academy of Sciences, as well as with a number of advisory committees within NASA itself, Fletcher became convinced that the development of the Space Telescope was timely and proper. The preliminary studies had shown that the telescope was feasible and that its cost would be manageable. Fletcher had at first been skeptical of the need for such a large telescope, and properly so, given his position as guardian of the reputation of NASA. In government administration as in personal life, no substitute exists for a high reputation, and NASA had acquired a dazzling one as a result of the successful Apollo moon landings four years earlier. Once he became convinced, Fletcher became a strong supporter of the Space Telescope project.

Richard Nixon was then president of the United States. His budget director scrutinized the spending plans for the Space Telescope in collaboration with his budget examiner for NASA, Hugh Loweth, and recommended it for a "phase B study," which means a complete engineering design. Given the expense of such a study, this meant that the administration was getting serious about the Space Telescope project. Astronomers were overjoyed, but they realized that only Congress can appropriate the money for government projects such as the Space Telescope. A majority of the 435 representatives and 100 senators had to be convinced of the telescope's worth.

The committee system in Congress makes this task less overwhelming than it sounds. Two types of committees in the House and two in the Senate, authorization and appropriation, have direct authority over NASA. In each house of Congress, there is an authorization committee that has the responsibility to oversee the NASA program (among others) and to assure that its proposals make sense. These two committees had generally been supportive of the NASA program, so it was not too hard for Fletcher to convince them to approve the Space Telescope.

The two appropriations committees, on the other hand, proved much more skeptical. The chairman of the Senate Appropriations Committee, William Proxmire of Wisconsin, had demonstrated a fondness for giving "Golden Fleece Awards" to recipients of federal funds whom he felt to be particularly undeserving, including one or two hapless scientists who were performing research that their colleagues had deemed worthy of support but that Proxmire in his wisdom regarded as a waste of money. The astronomers held their breath. Would Proxmire judge it a waste of money to spend a billion dollars to examine objects billions of light-years away from Washington, D.C.? Fortunately, the time of greatest danger passed, and Proxmire's committee signed off and gave its approval.

That left the House Subcommittee on Appropriations for Housing, Urban Development, and Independent Agencies, chaired by the redoubtable Hon. Edward M. Boland from the Second Congressional District of Massachusetts, whose Boland amendment prohibiting U.S. aid to the Nicaraguan *contras* was later to cause serious problems for Lt. Col. Oliver North. This subcommittee is responsible for overseeing NASA. Boland ruled his subcommittee in the congressional tradition, like a feudal prince. A Democrat from a safe district, Boland was assisted by Richard Malow, the key staff member for the committee.

Since the NASA budget was small compared to those of the Department of Housing and Urban Development and the Veterans Administration, both of which the Boland committee also oversaw, one might have assumed that Boland's committee had no time to probe deeply into the intricacies of a high-technology project such as the Space Telescope, which, even at the maximum rate of expenditure, would consume less than 5 percent of NASA's annual budget.

But the contrary proved to be the case. Between them, Boland and Malow had a detailed knowledge of the NASA budget, including the Space Telescope, and it soon became clear that Boland wanted to save the taxpayers money by not approving the Space Telescope. When the Boland committee report on the 1974 budget reached the floor of the House, funds for the Space Telescope had been deleted. Instead, the report contained a suggestion, not without merit, that NASA initiate a smaller project in space astronomy.

At that point, astronomers all over the country began to inform themselves about the process by which Congress passes legislation. Before long, informal lobbying for the Space Telescope took shape, complete with a phone campaign and similar tactics. Letters of support for the Space Telescope began to arrive in Washington from astronomers and their friends. The project became as visible as a proposed veterans hospital in some hospitable city. And veterans hospitals were things that Boland, as chair of the subcommittee that oversaw the Veterans Administration, understood.

During one of the many visits that a senior astronomer made to Washington to discuss the Space Telescope, he found himself in the office of a congressman (let us call him "A") who sat on the authorization committee for NASA. Suddenly the bell rang to signal an impending roll call on the floor of the House. Congressman A invited the astronomer to walk to the Capitol and to continue talking. As they were about to board the elevator, another congressman (let us call him "B") appeared. After friendly greetings, the three men entered the elevator. Talk turned to a favor that A wanted from B—a vote on a bill favorable to his district. B reminded A of a veterans hospital that he had been seeking for his district, and that needed approval from the authorization committee on which A sat. Congressmen A and B made a deal while riding in the elevator, and after B departed, A resumed conversation with the astronomer about the Space Telescope as if nothing had happened. Actually, the astronomer who had witnessed all this had received a valuable lesson in civics: if you want a congressman to support a project, you must demonstrate how it will benefit his or her district in a concrete manner.

The lobbying effort intensified. Senator Charles Mathias of Maryland became interested in the Space Telescope and succeeded in restoring the project to the NASA budget, although at a funding level only barely enough for planning efforts to proceed. Then, in August 1974, while these events were occurring, President Nixon resigned. Gerald Ford, who succeeded him as President, quickly put a hold on new projects. Even the small amounts already approved for the Space Telescope were cut in half.

NASA was also told to look overseas to find countries willing to contribute to the project. This was not at all unprecedented; NASA prided itself on the number of its projects involving international cooperation. But since the Space Telescope was complex and expensive, the negotiations took time. Ultimately, NASA did obtain agreement that the European Space Agency would contribute 15 percent toward the total cost of the Space Telescope project in return for 15 percent of the telescope's observing time.

By the time that this agreement was reached, two years had passed, and the Space Telescope once again came up for congressional review. In the meantime, Congressman Boland had stated, in a speech that he gave at the dedication of a ground-based radio telescope in Massachusetts, that (in effect) the Space Telescope would be approved over his dead body.

True to his word, Boland again deleted the Space Telescope from the NASA budget for 1976. This time, however, the supporters of the project in the House were ready, and when the NASA appropriations came up for a vote on the floor of the House, they moved to restore the Space Telescope. This motion carried, and so in 1977, several years after it was first proposed to Congress and thirty years after Lyman Spitzer first wrote about a large telescope in space, the Space Telescope project began.

But it faced one more funding hurdle. The relatively small amount proposed to be spent by NASA during the first year left room for the project to be killed the next year, when a major increase in funding would be necessary. This time, the problem arose with William Proxmire in the Senate. As a key committee chairman, his opinion counted, and the signs were not favorable. Again the astronomers sprang into action, and the Space Telescope project was saved in the Senate, largely through the efforts of Maryland's Senator Charles Mathias.

This time the victory stuck, and substantial funds began to flow to the project in 1978. NASA signed contracts with Lockheed for the spacecraft itself and with Perkin-Elmer for the telescope. The next seven years were good ones for the Space Telescope and led to its completion.

A multitude of problems did develop along the way: dust on the primary mirror, cost overruns on the scientific instruments, inflexibility in the operating software, difficulty in getting the fine guidance sensors to work properly, underestimates in the costs of future refurbishing by astronauts, inadequate provision for the ability of the telescope to track planets. The list could go on for a page, but eventually NASA and its contractors overcame each problem, although not without increased cost. Furthermore, all of these ground-based difficulties pale in comparison with the *Challenger* disaster in January 1986. The Space Telescope was to have been launched in October of that year, but the *Challenger* explosion delayed its launch by more than three years.

The most important lesson taught by the Space Telescope is this. Scientists dreamed of a new project. They proposed, worked, and lobbied for it. But in the end, the American people, through their chosen representatives, decided how to respond to the scientists' proposal. The result is not only a powerful new scientific instrument, alternately gleaming and hidden as it slips silently through space above the Earth, but a new and energized community of scientists, engineers, managers, and astronauts dedicated to the success of the American space program. Robert Wilson, Director of Fermilab, the great particle accelerator in Illinois, was once asked by a congressional committee how research in particle physics would contribute to the defense of the United States. "By helping to make it worth defending," Wilson replied.

3

WHY DO WE NEED THE SPACE TELESCOPE?

THE SPACE TELESCOPE represents the culmination of astronomers' dreams, the fulfillment of hopes borne for a generation or more, ever since the time that rockets first offered us the chance to place instruments above the veil of the Earth's atmosphere. Astronomers know better than anyone that we live on a speck of comic dust, adrift in a vast and nearly empty universe. On our planetary speck, itself accreted from materials formed over billions of years of time, we orbit our home star, one among hundreds of billions of stars in the Milky Way Galaxy. Astronomers dream of increasing our knowledge of the other specks in the universe, of solving the mysteries concerning their origin, composition, and ultimate fate. But why should the public care about fulfilling astronomers' dreams? What makes the Space Telescope so useful to us?

The short answer, of course, is that although the Space Telescope has little "practical" use, it will reveal startling new facts about the universe. Many arguments can be—and have been—made to justify projects like the Space Telescope to the public. Many of these arguments focus on the practical spin-offs of our scientific research, such as the no-stick Teflon that emerged from the early years of the U.S. space program. However, these arguments all suffer from being too roundabout a means to achieve a well-defined end. If we want a better no-stick surface for frying pans, our best move is probably to research frying pan surfaces, not heat-resistant nose cones that allow rockets to resist frictional heating in the atmosphere. Certainly no scientist would argue that the spin-offs from space research provide the chief reason to engage in such research. The real reason is simple: we do research because we are curious about the universe.

Since our resources are limited, part of our political life must resolve how much we are willing to spend to satisfy our curiosity. For many of us who are scientists, those amounts seem remarkably low in view of the fact that it must be our curiosity that allowed us to evolve to our present position as lords of the planet.

No one knows better than the builders of the Space Telescope that you cannot obtain government support simply because you promise to expand our knowledge of the cosmos. The Space Telescope exists because the U.S. Congress agreed with NASA that we should spend $2 billion to advance our understanding of the universe. To the extent that congressional action reflects the will of the American people, that means that we committed about $20 per family in order to place the Space Telescope in orbit. Why did we do this?

THE VEIL OF ATMOSPHERE

The fundamental reason for creating the Space Telescope resides in the air we breathe, our beloved, much-abused atmosphere, which is mostly a sea of nitrogen and oxygen molecules but contains additional molecules of water vapor, carbon dioxide, and other trace constituents, as well as an admixture of argon and neon atoms, inert gases that pass in and out through our lungs with no known effect whatsoever. This atmosphere, which was significantly altered by life on Earth long before human beings took their turn, is unique in the solar system; although other planets have gaseous envelopes, none has a nitrogen-oxygen shroud of gas. (The closest resemblance appears on Saturn's large satellite Titan, which has an atmosphere rich in nitrogen but with only trace amounts of oxygen.) The oxygen pumped into our atmosphere by plant respiration during the past few billion years allows oxygen-breathing animals to exist on the Earth's surface. But the oxygen has another role to play in keeping us alive: it helps to create the ozone layer.

THE GLORIES OF OZONE

Ozone molecules each consist of three oxygen atoms, held together by their mutual electromagnetic forces. Ordinary oxygen molecules, which we breathe every minute, consist of oxygen atoms joined in pairs (O_2), not in the triplets (O_3) that form ozone. In short, when it comes to breathing oxygen atoms, pairs are good, but triplets are useless or even harmful, because ozone molecules tend to annoy human throats and lungs. Once the ozone molecules do form, they have a particularly useful property unrelated to our respiration: they absorb ultraviolet light.

Ozone now receives much attention in the news because we are in danger of running out of it. Ironically, another problem with ozone is that we have too much of it. These apparently conflicting statements can be easily reconciled. We have too much ozone at low atmospheric altitudes, especially in cities, where automobile exhausts produce great quantities of ozone, and too little at high altitudes, where ozone proves useful in absorbing ultraviolet light.

One might think that this leaves us with no problem, since we can simply let the low-altitude, "bad" ozone rise to high altitudes to become "good" ozone. Unfortunately, this will not work. The ozone released at low altitudes can never rise to significant heights without combining with other molecules in the atmosphere and thereby losing its useful, ultraviolet-absorbing properties. Because this is so, our simplistic, intuitive approach to ozone cannot succeed.

To avoid the separate, harmful effects of ultraviolet light and of noxious ozone in the air we breathe, we must reduce our ozone emission at low altitudes, perhaps by designing more efficient automobile engines, and simultaneously end our depletion of high-altitude ozone. This depletion arises primarily from the release of chemicals called chlorofluoro-carbons, or CFCs, useful in refrigerators and in spray cans. A few steps toward this twin goal have been taken; far more will be needed. But if we hope to understand the full importance of ozone to us, we must acquire a better under-standing of the nature of light.

THE MANY FORMS OF LIGHT

Ozone molecules, as we have seen, absorb ultraviolet radiation. Ultraviolet radiation consists of waves that re-semble the light that our eyes see, but with slightly shorter wavelengths. All types of light—more technically, all types of electromagnetic radiation—consist of waves that some-how ripple through empty space. For any wavelike, rip-pling motion, the *wavelength* measures the distance be-tween successive wave crests. A mariner on a stormy day may encounter water waves whose wavelength is mea-sured in hundreds of yards, while a child in a schoolyard may see ripples in a puddle only an inch across. To an astronomer, the analog of the sea is empty space, and the waves ripple by themselves, quite capable of passing through a complete vacuum. In other words, light consists of ghostly waves, waves that require nothing to ripple through.

The wavelength of these ripples distinguishes one type of electromagnetic radiation from another. For visible light, we call the different wavelengths "colors." Red light is visible light with the longest wavelengths, and violet light is visible light with the shortest wavelengths; the other colors have intermediate wavelengths. Ultraviolet light consists of electromagnetic radiation with wavelengths slightly shorter than violet light, but unlike violet, ultravi-olet radiation is completely invisible to the human eye. Similarly, infrared radiation, whose wavelengths are some-what longer than those of red light, cannot be detected by the human eye, though our bodies detect infrared as a warm sensation, caused by the interaction of the infrared radiation with the molecules in our skin. Ultraviolet with wavelengths just a bit shorter than those of visible light

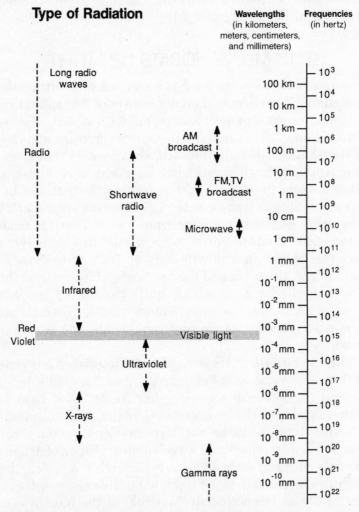

Figure 19: The different types of electromagnetic radiation, arranged according to their wavelengths or, alternatively, frequencies. *Drawing by Marjorie Baird Garlin*

leaks through the atmosphere and can cause sunburn in humans.

But ultraviolet and infrared fall far short of spanning the entire range of the possible wavelengths of electromagnetic radiation (Figure 19). At the very longest wavelengths, now-familiar radio waves have been used since the early years of this century to carry information around the globe. Between radio and infrared wavelengths lies the domain of "far infrared," still to be explored in detail by astrophysicists. At wavelengths shorter than those of ultraviolet, we encounter first x-rays and then gamma rays. X-rays are familiar for their ability to penetrate human flesh and thus to reveal the interior of the human body. Gamma rays, with wavelengths even shorter than x-rays, are associated with the transformation of one type of atomic nucleus into another, as occurs, for example, when radioactive uranium atoms spontaneously "decay" to form lead atoms. Such a nuclear transformation typically produces an accompanying gamma ray, which carries away some of the energy released in the transformation.

Most stars emit far greater amounts of visible light, ultraviolet, and infrared radiation than of the other types of electromagnetic radiation. Our sun, a typical star, follows this rule, so it is no accident that the first astronomers, scanning the heavens, noticed the stars and not the rest of the universe. But nature need not confine itself to visible light when electromagnetic radiation is emitted. Instead, as astronomers are now discovering, an abundance of objects emit one, several, or all of the types of electromagnetic radiation—both visible and invisible to the human eye.

Astronomers seek to detect and to analyze *all* types of electromagnetic radiation, in order to obtain the maximum possible knowledge of the cosmos. But a key limitation in the development of astronomy has been the fact that only a few "windows" in our atmosphere allow the penetration of electromagnetic radiation of certain wavelengths. These windows encompass the regions of visible light, along with some of the infrared and most of the radio waves. Other types of electromagnetic radiation simply cannot penetrate our atmosphere; instead, they are blocked by particular types of atoms and molecules at different altitudes (Figure 20). Hence only with the development of space technology,

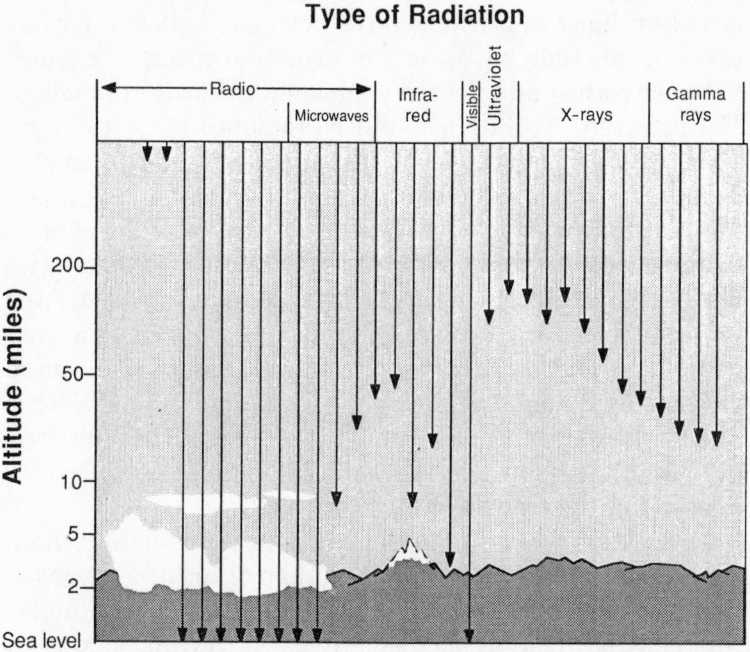

Type of Radiation

Figure 20: Atmospheric "windows" let visible light, some wavelengths of infrared radiation, and radio waves pass through to the Earth's surface, but other wavelengths of infrared radiation, as well as ultraviolet light, x-rays, and gamma rays, are absorbed by the atmosphere. *Drawing by Marjorie Baird Garlin*

which has allowed astronomical telescopes to orbit above the atmosphere, could astronomers begin to explore the universe using previously hidden regions of the spectrum of electromagnetic radiation.

The various wavelengths of electromagnetic radiation are correlated with the energy of the corresponding photons (units of radiation). Light and other types of electromagnetic radiation consist of massless particles called photons, which travel through space at the constant speed of 186,000 miles per second, the "speed of light." Photons have wavelike properties (for example, they each have a wavelength), but they have some of the properties of ordinary particles as well. In particular, each photon carries a certain amount of energy, despite having no mass. If a photon loses all its energy of motion, it simply disappears. Notice that although all photons travel at the same speed (the speed of light), they carry different amounts of energy. This may contradict our intuition, but is a fact of nature all the same.

Of the different types of photons, gamma rays have the highest energies, and radio waves the lowest. In other words, the photons' wavelengths and energies are inversely correlated; higher energies go with shorter wavelengths, and vice versa. We can now see why gamma rays, x-rays, and ultraviolet pose a danger to our health. These are the types of photons with the highest energies, and their impact on Earth poses the greatest danger to relatively fragile organisms such as ourselves. Since our atmosphere filters out all of these high-energy photons (except for some ultraviolet), human beings have evolved in comfort beneath its shield, rather than (as might have occurred) developing hard shells resistant to x-rays or finding habitats underground or deep beneath the ocean surface, where we would be safe if our atmosphere failed to block the high-energy photons. As events have unfolded, we live beneath a sea of air that is quite capable of protecting us from high-energy cosmic photons. And now that we have begun to master our environment, we know how to protect ourselves when we venture beyond our atmospheric shield. When we send a spacecraft or human being outside the atmosphere, we gain a crucial benefit: we can finally avoid the problems that the atmosphere brings to our study of the cosmos, even in visible light.

THE TWINKLING OF STARS

Anyone who has looked at the stars on a clear night has seen them twinkle, as each star appears to change rapidly in brightness. Examined with binoculars, the star also

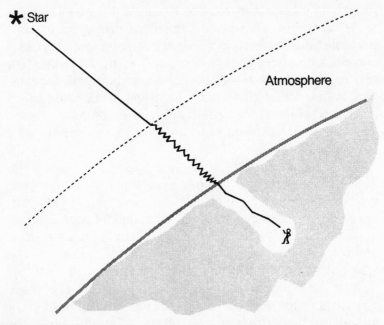

Figure 21: Twinkling arises from multiple refractions in the atmosphere as light passes through it. *Drawing by Marjorie Baird Garlin*

appears to change its position constantly. This twinkling arises not in the star itself, but far closer to us, in the Earth's atmosphere. As the beam of starlight passes through the air, it encounters thousands of individual parcels of air, which act as lenses. Each parcel bends the beam by a tiny amount (Figure 21). The net effect is that the beam is focused in a slightly different direction with a somewhat different brightness. As different parcels of air drift overhead, the star's image constantly changes in position and brightness.

The effect can be entrancing—or, if you are an astronomer, discouraging. You can never hope to hold the star's image exactly at a given point, nor to capture a perfectly focused view of a galaxy or a nebula, no matter how fine your telescope. Instead, the atmosphere will inevitably blur the image of the object you wish to study. Atmospheric blurring has two separate effects on your view of the universe, each of them capable of causing big problems for astronomers. These problems, interrelated yet fundamentally distinct, may be called the sharpness problem and the background problem. Because these problems underlie the need for the Space Telescope, they are worth examining separately and in some detail.

THE NEED FOR A SHARPER IMAGE

Astronomers and nonastronomers alike delight in a crisp, clear image of whatever they examine, be it the Great Red Spot of Jupiter or a photograph of a beloved grandchild. More than aesthetics is at work here. A sharper image contains far more information than a blurry image, and information is what we crave. A family photograph gains value in our eyes not so much for its suggestive power as for its informational content; if we see details such as the high school letters on a student's sweatshirt, these details add to our pleasure precisely because we do see them, not because we are led to imagine them. Similarly, the shape and shading of the image of a planet or galaxy can provide astronomers with important insights into the true nature of the object.

But nature imposes limitations on the sharpness of any image obtained by means of a system of lenses or mirrors, be it a human eye, a photographic camera, or an astronomical telescope. These limitations have a dual character, one technological, the other more basic. First, the sharpness of the image depends on the quality of the imaging system: Poorly made lenses yield fuzzier images than carefully ground lenses do. Our eyes function almost as well—for their size—as the highest-precision lens can, a tribute to nature's powers of creation. Because of the importance of lens quality, if we seek to improve our natural vision with binoculars or bifocals, it pays to obtain well-ground lenses and carefully mounted optical systems. Otherwise we are simply wasting our efforts to obtain a sharper image.

But even the finest optical system, whether natural or

human-made, has an unavoidable limitation on the sharpness of its images, a limit that arises from the laws of
physics. This limitation exists because light—or any other
type of electromagnetic radiation—consists of waves, ripples in the vacuum that always travel at the speed of light.
Any lens or mirror that forms an image does so by focusing
those waves to a particular point. This focusing can never
be perfect, because in addition to the waves that pass
through the center of the lens or reflect from the center of
the mirror, the lens or mirror also focuses waves that
encounter the edge of the lens or mirror. These edge-meet-
ing waves are imperfectly focused, because not only the
lens or mirror, but also the edge—which has no focusing
powers—acts upon them. As a result, the image formed by
the lens or mirror must always include imperfectly focused
waves. There is simply no way to eliminate this problem
completely, since every lens or mirror must have an edge.

What we *can* do is to make the lens or mirror larger. This
reduces the distortion of the image by reducing the fraction
of waves that meet an edge. This physical fact explains
why telescope makers have attempted to make larger and
larger lenses and mirrors for their telescopes.

If this were the entire story, by now we would have
optical telescopes as large as a football stadium. However,
two factors have kept telescopes from growing to such
extreme sizes under the pressure of our desire for sharper
images. First of all, it is difficult and expensive to make
enormous telescopes. Even more important is the effect of
atmospheric blurring. With a telescope smaller than a foot
or so in diameter, the effect is like what you would see with
binoculars—a single image of the star that changes constantly in position and brightness as different parcels of air
move in and out of the beam. But because parcels of air
tend to be about a foot across, the effect is different with
larger telescopes. Since beams of starlight reaching different parts of a large mirror arrive via different paths
through the atmosphere, they encounter different parcels of
air and are therefore deflected in different directions. If you

examine a star with a large telescope, you see not one, but many different images, each jiggling independently.

At a typical mountaintop observatory, the jiggling due to the atmosphere spreads the star image over about one second of arc, the size of a dime seen from a distance of two miles. While this is a sharp image by everyday standards, it is enough to reduce seriously the performance of large telescopes, to a level below the limits imposed by the laws of physics.

The designers of the Space Telescope knew that by placing a telescope above the atmosphere, they could eliminate atmospheric blurring. Theoretically, this would leave them limited only by the laws of physics and the fact that every lens or mirror has an edge. Then the largest possible mirror would produce the sharpest attainable image.

The designers therefore sought the largest telescope mirror that could be accommodated in the NASA budget. That constraint led to the choice of a mirror 2.4 meters (94 inches) in diameter. Astronomers would have dearly loved to build a telescope with an even larger mirror, but to place such a telescope into orbit would require more money than was available.

Nevertheless, the Space Telescope can produce images fifteen times sharper than those normally produced by telescopes at excellent sites on Earth. The Space Telescope will reveal details as small as one-fifteenth of an arc second, the angular size of a dime seen from a distance of thirty miles. Celestial objects have distances measured not in miles but in hundreds of millions of miles (for objects in the solar system), in many trillions of miles (for other stars in the Milky Way), or in millions of trillions of miles (for galaxies beyond the Milky Way). Even at such distances, the sharper images available from the Space Telescope are sure to reveal details never seen before, and therefore to show astronomers previously unknown aspects of the objects they love to study. For this result alone, the Space Telescope has been a dearly sought goal among astronomers for two decades.

THE SEA OF BACKGROUND LIGHT

Although it might seem sufficient for the Space Telescope to provide us with images fifteen times sharper than
any before available, a separate, related advantage of placing a telescope above the atmosphere is also significant.
That advantage becomes clear when we consider the second way in which atmospheric blurring hinders our vision
of the cosmos. Not only does it reduce the sharpness of an
image, but it makes it more difficult to pick out the image
against the background of other radiation. This second key
effect of atmospheric blurring on our search for astronomical objects seems a bit more subtle than the first; indeed, if
you talk to scientists in general, you will find that few of
them have thought much about it. The exceptions are
scientists, sailors, aircraft spotters, and others who have
tried to see a faint object against a bright background.

Consider the problem that a daytime airplane spotter
faced during wartime, before the installation of radar systems: You must find the aircraft as soon as possible, which
means looking for the sunlight reflected from the plane.
Your problem is increased by the fact that the sky is not
dark; instead, sunlight reflects from a host of molecules
and dust particles in the atmosphere, making the sky
appear blue, because the molecules and dust particles
reflect blue light more efficiently than other colors. When
you look for a distant airplane, you must recognize a faint
point of light not against a completely dark sky, but
against a sky that already sends plenty of light toward you.
This overall light from the sky forms what scientists call a
"sea of background light." Your task is not simply to spot a
faint image, but to discriminate between the light from that
image and the sea of light from the background (Figure 22).
This task is far more difficult than it would be if our sky
reflected no sunlight and therefore was dark even in the
daytime.

Now consider the effect that atmospheric blurring,
which changes continuously as our atmosphere ripples,
will have on this problem of background light. Suppose

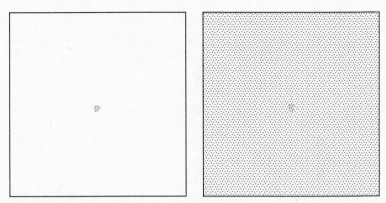

Figure 22: An object seen against a clear sky in the left panel is more difficult to see against the sea of background "noise" in the right panel. *Drawing by Marjorie Baird Garlin*

that the amount of sunlight reflected from the aircraft remains constant, but that atmospheric blurring spreads this light over an area much larger—say, ten times larger—than the aircraft itself. Then the maximum signal that your eye could receive from the aircraft's reflected light, now spread over ten times more area, would be only one-tenth as great as the maximum signal in the absence of atmospheric blurring. The signal of light reflected from the aircraft can therefore compete with the sea of background light only one-tenth as well. As a result, if the aircraft's reflected light would produce an image that lies just at the threshold of your detection capability *without* atmospheric blurring, then *with* blurring it would surely be undetectable. The spreading out of light by atmospheric blurring would make the object invisible, lost in the sea of background.

The same phenomenon operates with the night sky. Contrary to popular impression, the night sky is not completely dark, but glows faintly because of light reflected in the atmosphere from the moon and even the stars. Moreover, there is a faint emission (called airglow) from atoms and ions high in the atmosphere that are excited by energetic particles trapped in the Earth's magnetic field. And beyond our own atmosphere, there is a faint glow, called the zodiacal light, which consists of sunlight reflected from dust particles in the space between the planets. All of these sources of light contribute to a faint but appreciable sea of background light.

Suppose that an astronomer observes a globular star cluster, a collection of a million or so stars, and searches for the stars that are fainter than the few hundred brightest stars in the cluster. With a telescope on Earth, these few hundred are the only stars that can be seen; because atmospheric blurring spreads out the light of all the stars, the fainter stars remain invisible against the sea of background light. Now consider the same astronomer with the Space Telescope, observing the same globular cluster of stars. This fortunate astronomer profits in two ways from having a telescope in orbit above the atmosphere.

First of all, the sea of background light is less, because the telescope orbits above the region in the atmosphere that reflects moonlight or starlight. In addition, the Space Telescope orbits above the airglow and thus avoids it as well.

Second, and still more important, because the image of each star is much sharper, it is far easier to discriminate between the images of the stars on the one hand, and the background light that remains, even above the atmosphere, on the other. The background light consists of the zodiacal light and the light arising from the universe as a whole, representing the sum of all objects at all distances, most of them immensely great.

The combination of these two advantages enables the astronomer with the Space Telescope to detect stars with a brightness as little as one-fiftieth of that of the faintest stars detectable from ground-based observatories. This fifty-fold improvement should have a revolutionary impact on astronomical research.

THE SPACE TELESCOPE'S EFFECT ON ASTRONOMY

Whenever astronomers have used improved telescopes to look for fainter objects, they have found them, not in small but in enormous numbers. By allowing astronomers to detect objects only one-fiftieth as bright as the faintest objects they can detect now, the Space Telescope will permit them to find the same sort of object at a distance seven times greater than is possible now. (The apparent brightness of any object decreases in proportion to the square of its distance, and seven squared is roughly fifty.)

As a result, astronomers now believe that the Space Telescope will reveal a number of objects available for study that is not seven times, but the *cube* of seven, nearly 350 times greater than the present number. (The volume of space available for study, and hence the number of objects available for study under ideal conditions, increases in proportion to the cube of the distance to which we can see.) In short, take whatever host of objects you now imagine to exist in the universe accessible to our observation, and multiply that multitude by 350 to arrive at the host of objects accessible to the Space Telescope. As one astronomer put it with some hyperbole, "One picture with the Space Telescope is worth a thousand from the ground."

Perhaps even more important, the Space Telescope's ability to detect faint objects will undoubtedly reveal entire new *types* of objects, types now unknown and even unsuspected. This prediction may seem overly optimistic, but the rule has proved true each time a new technique for better astronomical observations has emerged. For example, only with the new, large telescopes constructed during the first half of this century could galaxies be recognized as true "island universes," collections of billions of stars. Before then, many astronomers believed that the spiral nebulae might be parts of our own Milky Way, which, they believed, might contain the entire universe.

In a way, this comparison with improved telescopes on Earth is too mild. The Space Telescope's improvement by a factor of fifty times in the minimum brightness of detectable objects over our best current telescopes represents the largest single leap in human detecting ability since the year 1610, when Galileo turned his one-inch telescope lens on the moon, the planets, and the Milky Way. Galileo's crude telescopes provided an increase in detecting ability over the unaided human eye by a factor of roughly 100. Galileo's observations changed forever the way we regard the cosmos; it is quite possible that the Space Telescope will do the same.

RESOLVING POWER: TO SEE TWO OBJECTS SEPARATELY

The twin improvements offered by the Space Telescope—sharper images and greater detectability of faint images—are impressive in themselves. We might take a moment longer to concentrate on an additional aspect of the first of these, another advantage that arises from a sharper image. A sharper image implies a greater ability to detect as individuals two objects that happen to be close to one another in space, rather than seeing them as a single, combined image.

Consider, for example, observations of the details in the atmosphere of the giant planet Jupiter. The images sent back to Earth from the two Voyager spacecraft in 1979 revealed that Jupiter's atmosphere has vortices, whorls, and filaments as fine in size as the spacecraft's cameras could detect (Figure 23). Meteorologists examining these photographs recognized at once patterns of motion reminiscent of the atmosphere of the Earth. The Voyager images whetted the meteorologists' desires to obtain images of these atmospheric features on a synoptic basis, that is, over a long stretch of time. The Voyager spacecraft, which passed Jupiter in a few hours, could not furnish such images. But the Space Telescope, even in orbit around the Earth, can see Jupiter's details almost as well as the Voyagers could, discriminating between a small whorl, perhaps, and another one that happens to be next to it.

One possibility for the Space Telescope that may not reach fruition—but a possibility fraught with implications—involves one of the oldest speculations: Are we alone in the universe? For the first time in history, the

Figure 23: The planet Jupiter, photographed from the *Voyager 1* spacecraft that flew past Jupiter on February 1, 1979. *NASA, Jet Propulsion Laboratory*

Space Telescope may allow astronomers to observe directly a true planet orbiting another star. Although astronomical speculation has been rife that such planets exist in abundance and may harbor forms of life something like our own, the astronomical evidence remains scarce. Until now, this lack of evidence has been unavoidable, because a ground-based telescope faces an impossible task in attempting to detect a planet around another star. The star shines a billion times more brightly than the planet does, since the planet reflects only the tiny fraction of the light of its parent star that it intercepts. Yet the distance between the star's image and the planet's image must be small, since the planet is in orbit relatively close to its parent star. Its simultaneous low brightness and proximity to a bright object make such a planet undetectable. Although various methods have provided indirect evidence that some planets exist beyond our solar system, other effects might yet explain the data, so doubt remains that planets have been detected by indirect means.

The Space Telescope, using what astronomers call the "coronographic finger," has a fighting chance to find planets—if they exist!—around any of the hundred or so stars closest to the solar system. The coronographic finger technique takes advantage of the fact that the Space Telescope produces such a sharp image of the star that we can effectively cover up that image, or most of it, and thereby detect objects close to the star that would otherwise be lost in its glare. With Earth-based telescopes, astronomers cannot effectively use this technique, because atmospheric blurring bends too much of the starlight around the finger. But in space, where no such blurring arises, astronomers can hope to block the starlight so completely that they may be able to spot a planet shining dimly by its reflected light, our speck-of-dust cousin in a star-dominated universe.

Equipped with instruments to take advantage of the absence of atmospheric interference, the Space Telescope will yield far sharper views of the cosmos in visible light, and far better images in ultraviolet light, than ever before available to us. Astronomers have waited so long for the Space Telescope—half of the working lives of many experienced scientists—that many of them have come to see it as the culmination of their hopes for space-borne instruments, the finest telescope they may see in their lifetimes. But before they can obtain new information about the universe, the astronomers must be sure that they know how to operate an automated observatory in space.

4

How to Operate an Observatory in Space

THE SPACE TELESCOPE, although aptly named, necessarily involves far more than a telescope and its associated equipment: It is an automated satellite in orbit around our planet, receiving commands, gathering data, and sending those data to Earth. (Once the Space Telescope has been deployed from the shuttle bay, the astronauts return to Earth, leaving the telescope in the automated mode.) The Space Telescope therefore requires the means of responding to the commands sent from Earth-based human beings that tell the telescope to seek, to find, and to observe particular astronomical objects. Complementarily, the Space Telescope must also store data temporarily and then send its results back to Earth for detailed analysis.

By the late 1970s, American engineers and scientists had acquired a great deal of experience in fabricating and guiding automated satellites, but the Space Telescope demanded solutions to much more difficult problems, chiefly because of the incredibly accurate pointing of the satellite that is required to take full advantage of the Space Telescope's ability to produce sharp images. These difficulties stretched the production time of the Space Telescope by several years, and although they were finally overcome, some compromises had to be made, fortunately without significantly affecting the performance of the instruments.

OPERATING AN AUTOMATED SATELLITE

If you want to use an automated satellite in orbit several hundred miles above the Earth, you must be able to send commands to the satellite and to receive data from it. The first order of business is to know where the satellite is.

Broadly speaking, all artificial Earth satellites move in one of two types of orbits. One is the synchronous orbit, in which the satellite circles the Earth at a distance twenty-two thousand miles above the surface of the Earth. An object in such an orbit completes each orbit in a period of one day, so if the satellite moves in the same direction that the Earth rotates, its motion synchronizes with the Earth's rotation. This means that if the satellite has an orbit above the equator and moves eastward, it *appears* stationary in the sky as seen from any point on the rotating Earth. If the satellite's orbit is inclined to the equator, but has the proper size for a synchronous orbit, the satellite will appear to move to the north and south of the equator, but its motion will still match the Earth's rotation. Anyone on the half of the Earth where the satellite is "up" (that is, the half of the Earth facing toward the satellite) can point an antenna toward the satellite and receive its signals indefinitely.

Most importantly, a satellite in synchronous orbit over the equator, although moving at nearly seven thousand miles per hour, appears to maintain a constant position in the sky. Here is the secret of television broadcasting via satellite: You, as the owner of a giant television network, send your satellite to a particular synchronous point, say, above the equator in Brazil, due south of New York City. (By the way, these synchronous points are starting to be rather crowded, so you had better act quickly.) Then your customers have an easy time in capturing your signal, once they have purchased a preprogrammed system to direct their satellite dishes toward the point above the southern horizon where your satellite is located.

If the Space Telescope had a synchronous orbit around the Earth, communicating with it would be simple—not quite as simple as that of a television addict with a satellite dish, but conceptually similar. Unfortunately, the Space Telescope cannot be sent up to a synchronous orbit. Unlike its smaller predecessor, the *International Ultraviolet Explorer* (which is in a synchronous orbit), the Space Telescope is simply too fragile to be launched on anything but the Space Shuttle, whose maximum attainable altitude with a full payload (and the Space Telescope represents a completely full payload) reaches no more than 375 miles above the Earth's surface.

Satellites like the relay satellites that will carry the Space Telescope's data to Earth ride the Space Shuttle to an altitude of a few hundred miles and are then launched much higher, into synchronous orbit, by their own, smaller booster rockets. This boost to higher orbit cannot occur for the Space Telescope, for two reasons. First, no adequate launch capability exists to boost an object as heavy as the Space Telescope out of Space Shuttle orbit more than twenty thousand miles higher, into a synchronous orbit. Second, in order to extend its useful lifetime, the Space Telescope system was designed so that it can be repaired and refurbished by astronauts in space. So long as the Space Shuttle provides the sole means of sending astronauts into orbit, they can rise no higher than 375 miles or so above the Earth. To send the Space Telescope higher than this, beyond the reach of adjustment or repair by human hands, would doom it to a shorter useful life.

COMMUNICATING WITH THE SPACE TELESCOPE

Because the Space Telescope orbits the Earth every ninety-six minutes, it quickly passes beyond the range of any ground-based tracking station. NASA must therefore communicate with the Space Telescope by using an intermediate series of satellites that are in synchronous orbit. (The old system of relying on several ground stations, spaced around the world, will soon be abandoned.) Although the method of operation is a bit roundabout, it should work perfectly well. No matter what position the Space Telescope occupies along its orbit, you simply try to make sure that somewhere above it—and not on the opposite side of the Earth—there is a relay satellite. Each relay satellite does move in a synchronous orbit and so can be easily tracked from a ground station on Earth. You need at least two relay satellites to assure that the Space Telescope almost always has a relay satellite in a direct line of sight, and that the relay satellite with which the Space Telescope communicates can in turn communicate with you (Figure 24).

So it will be with the Space Telescope. NASA's relay satellites are called TDRSS satellites (pronounced "teed-ress"), for the Tracking and Data Relay Satellite System of

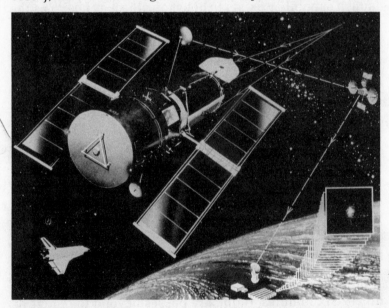

Figure 24: An artist's conception of how the Space Telescope sends data to a ground station on the Earth, relaying it via a TDRSS relay satellite. *NASA*

which they form an integral part. We may refer to them simply as relay satellites. The first flight of the revitalized Space Shuttle, the *Discovery* launch of September 1988, included the first of the TDRSS or relay satellites, which rode its own rocket from the altitude reached by the shuttle into the synchronous orbit path, at twenty-two thousand miles above the Earth's surface (Figure 25). The second relay satellite rode the third of the new Space Shuttle flights, once again in *Discovery*, in March 1989. With two relay satellites in orbit, NASA had achieved the basic condition for operating the Space Telescope, and will be in

Figure 25: A TDRSS relay satellite in the process of being deployed from the Space Shuttle. *NASA*

touch with it during 90% of each orbit. Without the relay satellites, launch of the Space Telescope would have been an exercise in futility.

GUIDING THE SPACE TELESCOPE SATELLITE

Picture an enormous tank truck, more than forty feet long and nine feet in diameter, moving in orbit at eighteen thousand miles per hour. Your job is to design a method to direct this twelve-ton machine by radio command, to point to a particular location on the sky, and to keep pointing to that location as the satellite moves in orbit with an accuracy so exact that if you are off by more than one hundred-thousandth of an inch, you have ruined your work. Furthermore, every few hours or so, you must arrange to swing this machine to another position on the sky, then to still another position, then to another, and so on—dozens of new pointings every day, many thousand in the course of a year of operation. How do you do it?

Astronomers know that the telescope must be pointed in stages. First you point the satellite toward a general location, and then, with fine guidance, you direct it accurately toward the particular object to be studied. To perform such pointing, most previous satellites used tiny jets from onboard rockets, but you can't do this with the Space Telescope. Even the small amounts of gas from such jets would contaminate the mirror and spoil your observations. So astronomers must rely on a new means of pointing the satellite, called reaction wheels.

Reaction wheels are large flywheels, mounted on bearings and constantly rotating, that can be linked to the rest of the satellite in one of two ways, either as an electric motor or as an electric generator. As a motor, the wheel imparts some of its rotational energy to the rest of the satellite. The satellite therefore tends to rotate in the same direction as the reaction wheel, which slows down slightly in compensation. In other words, the total rotational capac-

REACTION WHEEL ASSEMBLY

Figure 26: One of the reaction wheels that orient the Space Telescope.
NASA

ity (which scientists call the angular momentum of the system) remains the same, but the wheel has slightly less of it, and the rest of the satellite slightly more than before. As a generator, the flywheel reverses the process: The reaction wheel draws energy from the rotation of the satellite. This makes the wheel spin more rapidly, but the satellite spins more slowly (Figure 26). To use this method to point a satellite, you need three reaction wheels, one for each axis around which the satellite can rotate. (The Space Telescope has four wheels in order to provide some backup.) You also need a tremendously accurate system for control and feedback, in order to tell the operating system just what to do and for how long, so as to make the satellite point toward a particular spot on the sky.

Consider, then, what is involved in the motion of the telescope around a single axis. Guided by commands sent from Earth, the telescope makes an electrical connection to one of the reaction wheels, and the telescope begins to accelerate around that axis. Once the telescope has achieved the desired acceleration, the connection is disengaged, and the telescope rotates on its own momentum toward the chosen orientation. Then, as the telescope approaches the chosen orientation around that particular axis, the electrical connection must be reestablished, but in the opposite sense, so that the telescope slows down while the reaction wheel spins up, until the telescope coasts to a halt in its rotation around that particular axis at just the position desired when the commands were sent to the satellite.

Every swing of the telescope from one position to another therefore requires three separate phases—acceleration, coasting, and deceleration—around each of the three different axes. Nine different operations! Happily, the swinging around each axis can proceed independently of the other two, so that valuable time need not be wasted every time the Space Telescope is pointed toward a new object. Even so, each new object to be studied requires fifteen minutes or so simply to direct the Space Telescope toward a new point on the sky. Fifteen minutes per object, multiplied by a thousand different objects observed per year, adds up to a week and a half of lost time in every year of observation!

But it can't be helped. You can't observe a new object without swinging the telescope across a relatively large angle of the sky, and you can't swing the telescope at great speed, simply because the telescope has a large mass compared to the reaction wheels. (If you wanted reaction wheels as massive as the telescope itself, so that the fifteen minutes could be reduced to one or two minutes, you would need a Space Shuttle twice as powerful as the one we have to boost the additional weight into orbit.) The Space Telescope swings fast enough for its purposes; what counts most is its reliability (for it must swing from one object to another tens of thousands of times) rather than its speed.

The reaction wheels continue to play a role as the Space Telescope observes each particular object. Since the telescope tends to drift slightly as it observes a particular object, that object will quickly pass out of its field of view unless the position of the telescope can be stabilized by adjusting its orientation slightly every fraction of a second. This fine motion is likewise accomplished with the reaction wheels, but a tiny bit at a time. The telescope must be continually nudged so that the object it is observing won't stray from the center of its field of view. Otherwise the entire point of the Space Telescope would be lost; there is no reason to design and to launch the most accurate optical system in existence if you can't keep it pointed accurately at the desired object.

If the reaction wheels are functioning properly, it might seem a fairly simple matter to send commands from the ground stations to the telescope, using the Space Telescope's on-board computer, to point it toward a particular object in the sky whose position we know from ground-based measurements. In a broad sense, this assumption is true, but the method is nowhere near good enough. The Space Telescope has the most accurate view of the universe that any human-made instrument has ever achieved. If we simply used our previous maps of the locations of celestial objects, then we would be unable to fully use the Space Telescope's capabilities. That would be like directing a tremendously sharp knife while wearing thick gloves, because up until recently our knowledge of celestial positions has been crude by Space Telescope standards. The Space Telescope's ability to point at a particular object with incredible, never-before-achieved accuracy would be lost so long as we used only our previous knowledge of stars' positions to direct it.

Once the general area of observation has been set, astronomers require the Space Telescope to guide itself, finding and continuing to point exactly at the object desired. To do this, the Space Telescope must follow a two-step process: first establish a field of view, and then keep studying this field of view, not by commands from the ground, which would be hopelessly inaccurate, but by self-generated commands, as the Space Telescope continually notes its drift and corrects for it.

How can this be done? We need a guide star within the field of view, plus a feedback mechanism that will keep the telescope always pointed toward that guide star. The guide star must be a star whose offset from the desired object is known accurately, so while the telescope is oriented toward the guide star, the desired object will be exactly where it is desired in the field of view.

So when the Space Telescope achieves "first light," the primary worry of the astronomers and engineers who had coaxed the telescope into existence will be not so much whether the telescope would see light as whether the "fine guidance sensors" are working properly. The fine guidance sensors are the keys to the Space Telescope's successful operation, the most troublesome pieces of hardware aboard an exceedingly complex, occasionally cranky machine.

Three fine guidance sensors form the heart of the Space Telescope's pointing and control system, which probably posed the greatest technical challenge of all to the engineers who built the telescope. Each of the sensors detects light that has been deflected out of the main beam of incoming light toward that sensor. Once the telescope has reached the desired orientation, the on-board computer calculates the approximate position of two guide stars, relatively bright stars in the field of view. There had better be such stars; if not, the Space Telescope has no way to guide itself correctly to obtain a useful observation.

The fine guidance sensors must therefore be sensitive enough that any random field of view has a good chance of including two stars bright enough for the sensors to detect. Since the average field of view contains no truly bright stars, nor even any rather bright stars, this requirement means that the fine guidance sensors must be capable of routinely detecting faint stars, which are the only stars available to serve as guide stars in a typical field of view.

Using its fine guidance sensors, the Space Telescope must play the game of "find the field of view" *every time* that it seeks to make an observation. First, relying on gyroscopes that describe the approximate orientation of the spacecraft, the Space Telescope swings itself into the approximately correct direction. Then fixed-head star trackers, which are essentially small telescopes with a relatively wide field of view, look for rather bright stars whose positions have already been fairly well determined from the ground. These stars' positions are already stored in the on-board computer. Once the fixed-head star trackers have located the bright guide stars that should be in their wide field of view, the fine guidance sensors seek other, much fainter stars, which will serve as reference objects within the much smaller field of view of the Space Telescope itself.

To do this, the computer figures out whether or not the fine guidance sensors ought to find a star within the field of view that is bright enough to make the sensors respond when the starlight falls on them. One fine guidance sensor begins to search for the first guide star, moving itself slightly to cover its field of search in a tight spiral pattern. Once this fine guidance sensor detects a suitable star, it stops searching; the other fine guidance sensor then begins to search for *its* guide star. When the two guide stars have been found and their positions measured according to ground-based measurements, the distance between the two stars on the sky is checked to see if it equals what it should. If it does, chances are that the pair of stars is the correct one, and the search is terminated. If not, the search starts all over.

This is not a simple affair. Each search typically requires about ten minutes—ten minutes that could otherwise be used for observing the desired astronomical object. And these ten minutes elapse only *after* the Space Telescope has been swung into the approximate position, a process that itself requires some fifteen minutes! There is no better way to operate; if you want to use a telescope with an incredibly sharp view of the universe, you have to be ready to take the trouble to point it with incredible accuracy. Once the telescope is properly pointed, the fine guidance sensors continually send information to the computer, which in turn tells the telescope whether or not to lock on to a given reaction wheel, and whether to accelerate or decelerate around a particular axis—an ongoing, never-ending, complex operation that must occur before any observation can have the least chance of success.

When the fine guidance sensors detect their two guide stars, the stars' images are directed toward Koesters prisms, glass wedges that divert the two beams of light and keep track of their location, so that the telescope knows whether or not it is continuing to point at a particular place and to receive light from the guide stars. When the system works—and it must work!—the entire twelve-ton instrument can be continuously pointed with an accuracy of at least 0.012 second of arc and (astronomers hope) possibly to an accuracy of 0.007 second of arc—equivalent to holding the telescope steady to a hundred-thousandth of an inch!

There is one more problem, though. The entire pointing operation depends on finding two guide stars in every field of view and on using the positions of those guide stars to direct the telescope toward the object in the field of view that it is to study. This requires that every field of view possess two guide stars bright enough to make the fine guidance sensors respond, with positions accurately known so that these positions can be used to direct the telescope. Until recently, astronomers simply did not know sufficiently accurate positions (0.3 second of arc or better) for enough stars to assure that each field of view will contain two suitably measured guide stars.

Hence one of the tasks of the Space Telescope Science Institute, which directs the Space Telescope's scientific research, was to establish a catalog of guide stars whose positions were to be stored in the on-board computer. To do this required measuring with the greatest possible accuracy the apparent positions of millions of stars on photographic plates of the sky obtained by the forty-eight-inch Schmidt Telescope at Palomar Mountain in California, and at a similar telescope in Australia that observed the southern skies. Even though automated, this task took three years to perform. With a new catalog containing 20 million star positions, the Space Telescope Science Institute fulfilled its mission, so that the Space Telescope could avoid the disgrace of finding itself all launched up with nowhere to point.

Once the Space Telescope points itself properly, its true function—observation of the cosmos—can begin. But it is not enough simply to begin; the satellite must continue to point, with the same incredible accuracy, at its observational target during its entire observation. Otherwise the entire effort of placing a telescope in a position to obtain the sharpest possible views of the universe will have been wasted. Successful operation of the Space Telescope therefore includes the requirement that a pointing accuracy better than 0.012 arc second be maintained for half an hour or more, until the telescope swings toward another target— which in turn requires the same pointing accuracy for the same amount of time.

The fine guidance sensors that find the guide stars in the field of view allow the Space Telescope to track these guide stars, using a feedback loop that sends information about the location of the guide stars to a set of gyroscopes, and updates the information once per second. The gyroscopes in turn tell the spacecraft whether or not to engage its reaction wheels to correct for any small deviation from the correct position. Tests indicate that all of this should proceed quite smoothly, so long as the telescope does not undergo any sudden jitter that would carry the guide stars out of the field of view of the fine guidance sensors. The Space Telescope scientists refer to this possibility of jitter as "losing lock" on the guide stars. If that occurs, the procedure of finding the guide stars must begin all over again, with the loss of valuable minutes of observing time.

What could cause the telescope to lose lock on the guide stars? The most likely offenders are motions of individual components within the Space Telescope itself. Because the angular momentum of the telescope as a whole must be conserved according to Newton's laws of motion, such motions will react on the overall optical system of the Space Telescope, causing it to move away from the direction in which it is pointed. For example, when one or more of the filters that form part of the spectrograph are inserted into or removed from the beam of light, or when the reaction wheels engage or disengage with too much of a jerk (actually the slightest sort of nod), the fine guidance sensors may lose sight of the guide stars. Computer simulations and ground tests suggest that such lock loss will occur only rarely, but experience in space will tell the full story.

For now, the Space Telescope is ready to track stars accurately enough to point the telescope at one automobile in a parking lot, and not at its neighbor, if the telescope were observing a parking lot on the moon! If we can do that, who knows what else is possible? To answer that question, we must examine the business end of the Space Telescope: the scientific instruments that it carries.

5

THE SCIENTIFIC PAYLOAD

THE SPACE TELESCOPE satellite represents a marvel of engineering, a telescope platform that can be pointed with greater accuracy than anything that humans have previously achieved. But the astronomical focus of attention—and the spacecraft's reason for existence—lies in its scientific instruments, in its ninety-four-inch reflecting telescope and the detectors that analyze what the telescope sees. Everything that has come before—every false start, every congressional briefing, every engineering hurdle, every problem solved in mating the satellite to the Space Shuttle and the telescope to the satellite—has had a single purpose in mind: to place above the atmosphere a large, automated machine that can detect and analyze light far better than anything on Earth. And for that, you start with a telescope.

WHAT *IS* A TELESCOPE?

A telescope owes its name to the Greek words for *far* and *see*, but this name actually describes what a telescope offers its users, not what it *is*. Scientifically speaking, a telescope is an object that produces an "image" and studies that image.

And what is an image? An image, whose name reminds us of its not-quite-real, quasi-imaginary existence, is an optical reproduction of reality. The simplest type of image is one formed by a single lens. Consider, for example, a child who uses a magnifying glass to project an image of the sun onto a piece of paper. The lens focuses the sun's rays of light by bending each of them (save the one that passes directly through the center of the lens) by a small amount. As a result of this focusing, the lens produces an image—a concentration of light waves that exactly reproduces the appearance of the object whose light passes through the lens. The exactness of the reproduction arises from the fact that the light rays are each bent according to the same rules. In practice, no image can be a truly exact reproduction, because no lens is perfectly formed to bend light according to a perfect set of rules.

The image of the sun produced by a small magnifying glass has a diameter of less than one-tenth of an inch and therefore looks like a single point. (As every Boy or Girl Scout knows, this point concentrates the sun's rays enough to ignite a paper if the light shines steadily upon it.) If you want to create a larger image of the sun, you need a lens with a longer *focal length*—one whose image is created at a greater distance behind the lens. Typically, the larger the lens, the larger the focal length (which measures the distance from the lens to the image that it forms) and the larger the diameter of the image. For this reason, photographic telephoto lenses tend to be wider as well as longer than ordinary lenses.

A telescope with an effective focal length of eight or ten feet will produce an image of the sun about an inch across. If you take the eyepiece lens off such a telescope, you can use the telescope's large ("objective") lens to project an image of the sun onto a piece of paper (Figure 27). This image will be large enough to reveal sunspots and other features on the solar disk. Both of the authors of this book performed this feat with small telescopes as part of their early education in astronomy, and were thrilled with the result. The laws of optics worked, and the image of the sun suddenly appeared, displayed on the paper held behind the telescope.

Figure 27: Looking at sunspots: a small telescope with a large effective focal length is being used to project an image of the sun onto a piece of paper. *James Cornell, Smithsonian Astrophysical Observatory*

The sun's image is the only astronomical image that is large and bright enough to be projected directly, so that it can be examined without using a second, eyepiece lens on a telescope. In most situations, the telescope creates an image that is so faint and so small that we can examine it only with the help of a second, smaller lens that magnifies the image. An astronomer who seeks to study a particular object chooses an eyepiece lens that provides the degree of magnification desired, always conscious that atmospheric blurring makes too much magnification useless.

We can obtain some useful insights into imaging the universe by comparing a telescope with a human eye. The eye has a lens analogous to a telescope's objective lens, which produces an image by bending the rays of light that

pass through it. This image falls on a light-sensitive surface, the *retina* of the eye. Notice that the eye makes do
with a single lens; the human body has no way to magnify
the image once the eyeball forms it. Therefore we must use
binoculars and telescopes outside the eye to achieve such
magnification.

In a marvelous display of organic chemistry, each part of
the retina responds to the light that reaches it with chemical changes that the light causes in a type of molecule
(called rhodopsin) abundant in the retina. As a result of
these changes, any image that falls on the retina registers
itself there through the workings of rhodopsin. But if we
want this process to be of any use to us, we need a way to
"read out" the image or, more precisely, to interpret the
effect that the image of incoming light has produced on the
retina. This readout mechanism consists of the nerve endings that are embedded in the retina and that can sense the
chemical changes that occur there, coupled with the nerve
fibers (the optic nerves) that pass the information about
the changes from these nerve endings to the brain. Our
brain knows what to do with the information. Somehow
(we don't know exactly how) it processes the arriving data
and creates within us (but just where is a matter of some
speculation) a picture that corresponds (so we think!) to
the actual world.

In this comparison, the lens of the human eyeball corresponds to the objective lens of a telescope. In the case of the
Space Telescope, it corresponds to the ninety-four-inch
mirror that focuses the light to form an image. The retina of
the human eye corresponds to CCDs (charge-coupled devices), light-sensitive electronic devices that register the
impact of light by creating an electrical current. The optic
nerves correspond, in the case of a CCD detector, to the
wires that carry the responses from the CCD to a computer
for processing, and the human brain corresponds to the
central processing unit of that computer, which creates an
image for later display on a television monitor or for magnetic storage on disk or tape.

THE MANAGEMENT OF LIGHT IN THE SPACE TELESCOPE

When scientists and engineers came to design the Space Telescope, much of their planning was driven by a key technical fact of life. The CCD detectors that play such a critical role by sensing the light that reaches them and by transmuting their detection of light into electrical impulses are limited by the size of their component parts, called pixels (short for "picture elements"). Each pixel in a CCD detector forms an element of the mosaic that detects and reproduces an image; a pixel is the smallest unit of the detector that can sense an incoming photon, and thus is analogous to one of the rods or cones of the human retina.

In theory, astronomers would like the pixels to be as small as possible, in order to obtain the sharpest possible image, one that consists of as many individual elements as possible. Our eyes contain thousands of rods and cones, rather than a few dozen, because we need a relatively sharp view of our environment. But this theoretical requirement must confront the fact that it is useless to provide more pixels than can be matched to the focal length of the Space Telescope's mirror.

To obtain a sharp image from the Space Telescope or any other optical system, you seek as long a focal length as possible, in order to produce as large an image as can be obtained, and the smallest pixels that can be produced to see that image as clearly as possible. The ideal combination would be a telescope with an infinitely long focal length and detectors with pixels infinitesimal in size. This situation, however, cannot be realized within the NASA budget.

Modern technology can make pixels that are marvelously small, only one two-thousandth of an inch in diameter, so that many pixels can be put into a practical detector. The CCD detector on the Space Telescope includes 2.56 million pixels. To make the fullest possible use of the amazing sharpness of the image that the Space Telescope produces, the size of the smallest detail discernible in the image should match the diameter of the pixel. If the image details were smaller than the pixel size, some of the image sharpness would be lost in the detection process. If the details were larger than the pixel size, there would be no point in having made the pixels so small. The requirement that the pixel size should match the size of the smallest details in the image implies that the telescope's focal length should be 190 feet. A shorter focal length would lose the chance for the sharpest possible image we could detect, whereas a focal length longer than 190 feet would be foolish, since the image would then be even sharper than the pixels could detect.

The Space Telescope scientists and engineers were constrained by the fact that in order to fit inside the Space Shuttle, the telescope could be no longer than forty-two feet. In the simplest kind of reflecting telescope, the mirror focuses the incoming light to a point at the other end of the telescope. In such a telescope, the constraint imposed by the shuttle's size would have limited the telescope's focal length to less than forty feet.

However, wonderful tricks are available to those who know how to build telescopes. As Figure 28 shows, the Space Telescope includes not only a large primary mirror, 94 inches in diameter, that focuses the incoming light, but also a secondary mirror, 12.5 inches across, supported sixteen feet ahead of the primary mirror by a truss that projects into the path of incoming light but blocks only a small portion of that light. Incoming light reflects from the primary mirror toward the secondary mirror, which in turn reflects the light through a small hole in the center of the primary mirror, where the light can be detected and analyzed by the instruments. The actual image produced by means of this double reflection is located a few feet behind the primary mirror, and the total effective focal length of the telescope equals 190 feet.

In other words, optical techniques lengthen the path of light reflected by the primary mirror as it forms an image, enabling the focal length to increase by a factor of four over

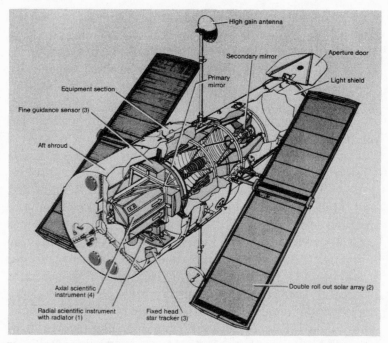

Figure 28: Cutaway diagram of the Space Telescope. *The Space Telescope Science Institute*

what it could be achieved with a single reflection. The size of the image likewise grows four times larger than it would be if formed through the single reflection from the primary mirror. Astronomers call this situation of multiple reflection and increased focal length a Ritchey-Chrétien variant of the Cassegrain optical system. The names commemorate scientists who designed improved types of telescopes, and the Space Telescope represents the culmination of nearly four centuries of telescope design.

The Space Telescope has a field of view that is nearly half a degree across, about the angular size of the full moon as seen from Earth. If the telescope were to study the moon (but it never will—the moon is so bright that its light would damage the CCD detectors), the moon's image would be twenty-three inches across. Note that the full moon, like the sun, spans thirty minutes of arc (half a degree) on the sky. A full circle around the sky spans 360 degrees, so it would take 720 suns or moons, standing shoulder to shoulder, to circle the sky at the horizon. To an astronomer, however, the thirty minutes of arc covered by the sun or moon represent a huge fraction of the sky. Astronomers often deal with objects that span no more than a few *seconds* of arc. Each second equals $1/60$ of a minute of arc, and thus equals $1/3,600$ of a degree of arc.

The human eye sees the world with a sharpness of about 1 minute of arc; if two objects are located closer than this on the sky, they appear as a single object to the unaided eye. This 1 minute of arc is the *resolving power* of the unaided eye—the measure of its ability to discriminate between two sources of light spaced by a small angular distance. Galileo's primitive telescope had a resolving power of about five seconds of arc (one-twelfth of a minute of arc); thus it allowed a discrimination between nearby objects about a dozen times finer than that of the eye alone. The Space Telescope will have a resolving power of 0.06 second of arc—nearly a hundred times better than Galileo's telescope and a thousand times better than the resolving power of the human eye (Figure 29).

WHY CAN'T WE SEE WITH INFINITE SHARPNESS?

All of us know that if we try to read smaller and smaller print, we eventually reach a size for which we simply cannot distinguish the individual letters. Unfortunately this problem, and similar ones, grows steadily worse as we grow older. But even the most visually adept—the Ted Williamses and George Bretts among us—cannot see with perfect clarity; there is some print too fine for anyone to read. A bit of thought about why this is so highlights the importance of the Space Telescope.

Our ability to see detail depends on two characteristics of the human eye: the focal length of the eye's lens and the pixel size of the detectors (rods and cones) in the retina.

M51 AT 1 AND 30 RESOLUTION

Figure 29: The spiral galaxy M51 photographed with a resolving power of thirty seconds of arc *(right)* and one second of arc *(left)*. The improvement by a factor of thirty in resolving power is approximately the factor by which Galileo's telescope improved over the human eye, and about twice the factor by which the Space Telescope will surpass the best ground-based telescopes in resolving power. *NASA*

The lens's focal length determines how carefully that image is scanned, bit by bit. The lens of the human eye has a focal length of about half an inch, and it produces an image of our field of view about one-quarter of an inch in diameter. Within this image size, many thousand rods and cones are arranged in an array that we may simplify by imagining it to consist of 500 by 500 individual detectors. If we want eyes that see more sharply, we need eyes whose lenses have a longer focal length, and more rods and cones (more pixels) to read out the larger image that the lens produces.

Since we don't have the time to let nature evolve this system—even if evolutionary pressures are pointing in this direction—we can use artificial eyes. The Space Telescope represents the latest and best of these: it has a focal length about five thousand times longer than the focal length of the human eye and therefore produces an image spread over about five thousand times the diameter of the image in the human eye. In principle, then, the Space Telescope could accommodate vastly more pixels in its detectors than the human retina does, but with our present technology, the Space Telescope must make do with "only" 2.56 million pixels in the CCD detectors of its chief camera. But even this "modest" number of pixels will allow the Space Telescope to vastly improve our ability to read the fine print of the cosmos.

WHY DO WE CARE ABOUT LIGHT-GATHERING POWER?

The ability to gather light, and thus to see objects even with extremely dim illumination, has played a role of differing importance for different species of animals on Earth. For those who live by hunting prey at night, such as the cat family, the need for extra light-gathering power has been crucial. Such animals therefore tend to have large eyeballs, with pupils that can open completely to take full advantage of the large size of the eye. This allows the predators to see objects in light too faint for smaller-eyed prey to see anything.

Light-gathering power is distinguishable from resolving power, though the two often go hand in hand in bigger and better telescopes. The laws of physics require that if you want high resolving power, the edges of your lens must be as far apart as possible, and therefore you need a large diameter. But since the parts of the lens or mirror that are relatively close to each other—that is, the parts near the center—don't contribute much to improving resolution, you might as well discard them, leaving only a ring at the edge of the lens to focus the light. Indeed, high resolving power can be achieved with just a few pieces from the edge, perhaps just one from each side, to produce the image that you seek.

This is the principle behind the interferometer, a special type of telescope made with only a few pieces of the reflector. (Mirrors rather than lenses are used in almost all interferometers.) Suppose that you take two mirrors and regard them not as separate but as the opposite edges of an imaginary, much larger mirror. If you combine the images from the two mirrors, using advanced optical techniques, you can obtain the much greater resolving power of a single larger mirror, one whose diameter equals the spacing between the two components. You have not gained much in light-gathering power (only twice that of each of the mirrors), but the resolving power is much greater than that of either mirror.

Although the first interferometers used optical mirrors, today the great use of interferometers occurs in radio astronomy. Radio waves have much longer wavelengths than visible light, and the laws of physics imply that radio waves are correspondingly far more difficult to bring to a sharp focus. For that reason (among others), radio telescopes have diameters measured not in inches but in many feet; the largest of them, the giant Arecibo radio dish in Puerto Rico, spans 1,000 feet, sunk into a natural limestone bowl near the city of Arecibo. Such radio telescopes do not form an image that we can see; instead, the radio waves, focused by the parabolic reflector, are analyzed electronically. But the same rule that applies to focusing light waves applies to radio waves: if you want a sharper image, you need a larger telescope—or at least pieces of one.

So radio astronomers began to build radio interferometers, arrays of radio telescopes spaced at intervals of thousands of feet. The radio signals detected by each telescope are fed into a central computer, which combines and processes them, treating the telescopes as pieces of a single, imaginary, giant dish. Such a dish can achieve far greater resolving power than any single radio telescope can. These interferometers have worked wonderfully well, and they provide our views of the universe with the highest resolving power now available. The most impressive sight in the world of interferometers rises from the plains of San Augustin in New Mexico, where twenty-seven radio telescopes ride on railroad tracks, movable to different positions (to add detail to our view). All twenty-seven dishes, which spread over a Y-shaped system that branches out for a dozen miles in each direction, function together, providing a resolving power equal to that of a single radio telescope two dozen miles across!

The Very Large Array (VLA) hardly ends the quest for resolving power among radio astronomers. By now they have arranged for telescopes at opposite ends of the United States—indeed, at opposite ends of the Earth—to be linked together electronically, to furnish a system with a resolving power equal to a telescope 8,000 miles in diameter! Further than this we cannot go—until we put radio telescopes in space. The interferometer method of increasing resolving power is far easier to implement for radio waves than for visible light and ultraviolet. For this reason, our attempts at increasing the resolving power of "ordinary," optical telescopes still rest with making larger mirrors and sending them above the atmosphere.

In contrast to a telescope's resolving power, which we can increase by constructing small pieces of an (imaginary) large mirror, the light-gathering power depends on gathering all the light possible—all the light that falls on an entire mirror. Light-gathering power, straightforwardly, refers to a telescope's ability to gather light waves. Here there is no quick fix: If you want to gather lots of light, you need a single large mirror to catch it. You can't use pieces of a mirror because they will gather only a tiny fraction of the amount of light that the full mirror would. Hence an interferometer proves useless for observations of an extremely faint source of light; the interferometer simply cannot trap enough photons to allow observers to distinguish the object against the sea of background radiation. There is no point in training a telescope with tremendous resolving power on an object that it can't see!

Some astronomical objects are bright enough that light-gathering power is not a problem; even the relatively low light-gathering power of an interferometer will suffice. But for the faintest sources of radiation, we must struggle simply to detect the sources, and must worry less about bringing fantastic resolving power to bear. For faint sources, we simply want as large a mirror as we can deploy.

Nature has never developed the interferometer. Our eyes and brains do not combine the images that the two eyes receive in a way that would allow us to resolve greater detail in the images. (If they did, we would notice that the world looks much blurrier when seen through a single eye than when we observe with both eyes.) Nature has taken advantage of the principle of range finding, so that our brains combine the images from both our eyes to estimate distances far better than we can by using a single eye. However, an interferometer does much more than this: it makes two or more detectors truly work in tandem, functioning as pieces of a single, larger detector. Hence, in nature, greater light-gathering power, which depends simply on the size of the eye, typically marches step by step with greater resolving power. A hawk or a tiger can see finer detail (greater resolving power) and can also see in fainter light (greater light-gathering power) than an animal with smaller eyes can.

Like the eye of a hawk, the Space Telescope has a single light-gathering component, not the several pieces that are found in an interferometer. The Space Telescope's ninety-four-inch mirror represents a compromise, not as large as the one that astronomers originally wanted, but large enough to provide great light-gathering power, especially since the background light will be far less in space, and the sharpness of view will help the telescope to distinguish objects against the background. Furthermore, the Space Telescope's resolving power, although not as great as that of a larger telescope, will represent a tremendous improvement over any Earth-based optical telescope, simply because atmospheric blurring will be absent. Indeed, since the Space Telescope's guidance system functions at the limit of our technological ability, we would probably be unable to profit now from the greater resolving power of a significantly larger telescope in space. We could, however, use its greater light-gathering power to observe even fainter objects than the Space Telescope can.

THE ARRANGEMENT OF INSTRUMENTS ABOARD THE SPACE TELESCOPE

Picture the Space Telescope in orbit, pointed in a chosen direction by its reaction wheels, which function in collaboration with the fine guidance sensors and the gyroscopes that learn from the sensors every second just where the telescope points. Carefully prevented from pointing anywhere close to the sun, the Earth, or even the moon (whose brightness would ruin the delicate instruments), the telescope directs its attention toward the light from, say, a distant galaxy. The galaxy's light may have traveled for five billion years to reach the Earth, so long that the galaxy may have burned out in the meantime. Finally, a tiny portion of that light passes into the opening at the front end of the Space Telescope, an eight-foot aperture designed to gather light, sort it into its components, and yield more information about the universe than one can easily believe possible. Just how does this information emerge?

After the light is focused by the primary mirror and then by the secondary mirror down through the small hole in the primary mirror toward the instruments, part of the beam of light is "picked off"—reflected—by one of the four pick-off mirrors that redirect the incoming beam toward different locations behind the primary mirror (Figure 30). Three of these pick-off mirrors are used to direct light to the fine guidance sensors that help to point the telescope. The fourth pick-off mirror directs light toward the wide-field/planetary camera, which secures wide-field images of whatever the Space Telescope is studying. We may simplify the camera's name and call it the "wide-field camera."

The parts of the beam of light that are not intercepted by one of the four pick-off mirrors pass onward toward four

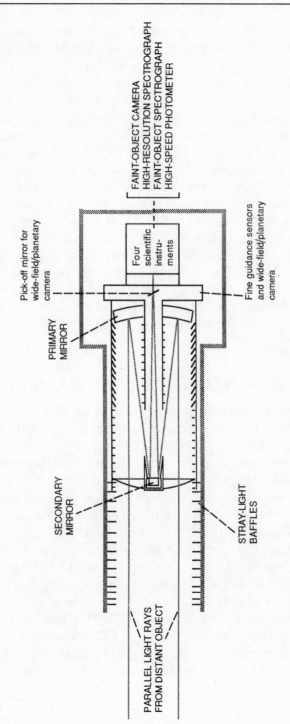

FAINT-OBJECT CAMERA
HIGH-RESOLUTION SPECTROGRAPH
FAINT-OBJECT SPECTROGRAPH
HIGH-SPEED PHOTOMETER

Pick-off mirror for
wide-field/planetary
camera

Four
scientific
instru-
ments

Fine guidance sensors
and wide-field/planetary
camera

PRIMARY
MIRROR

SECONDARY
MIRROR

STRAY-LIGHT
BAFFLES

PARALLEL LIGHT RAYS
FROM DISTANT OBJECT

Figure 30: The path of light in the Space Telescope, showing the location of the pick-off mirror for the wide-field/planetary camera.
Drawing by Marjorie Baird Garlin

key instruments. Each of these instruments—one camera, two instruments to analyze the colors of light, and one instrument to measure the total intensity of light—can study the light reflected by the primary mirror when the telescope is pointed so that the target's image falls on the entrance to the instrument. At any given time, each of the five instruments receives some of the light passing through the mirror from some part of the field of view. But because interesting astronomical objects are usually found in complex random patterns on the sky, if the light from a particular object falls on the entrance to a particular instrument, there will be little of interest falling on the other instruments, with the possible exception of the wide-field camera. Since the entrance to each instrument fills only part of the field of view, the commands sent to the Space Telescope must assure that the light from the object to be studied by a particular instrument does reach the entrance to that particular instrument within the field of view (Figure 31).

THE WIDE-FIELD/PLANETARY CAMERA

Of the five major instruments aboard the Space Telescope, the wide-field camera is the easiest to understand. This camera continuously records the image of whatever falls on its entrance, no matter what the other instruments are doing. Like the retina of the eye or the film in an ordinary camera, the camera detects photons focused onto it. The heart of the wide-field camera is the CCD detector made of 2.56 million pixels. When light strikes an individual pixel, that pixel produces an electrical charge in an amount proportional to the amount of light striking it within a preset interval of time, which may vary from a few seconds up to half an hour.

To reduce stray electronic noise in the detectors—what scientists call the "dark current"—the CCD detector must be kept at a temperature of −140 degrees Fahrenheit. This requires a cooling system, which draws its power from the

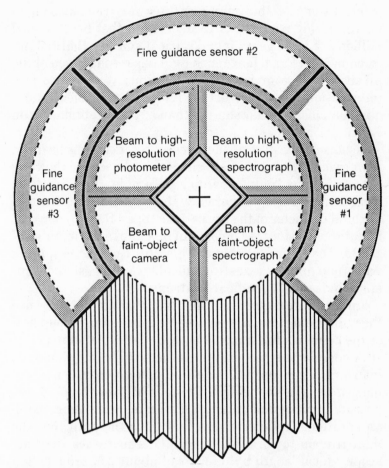

Figure 31: The arrangement of the four focal-plane instruments behind the primary mirror of the Space Telescope. *Drawing by Marjorie Baird Garlin*

solar panels and storage batteries, because the temperature of the telescope in Earth orbit averages about 50 degrees Fahrenheit.

Part of the CCD detector reads out the amount of charge that has accumulated at each of the pixels and sends the total readout to the on-board computer for storage. Every half hour or so, the computer sends its accumulated information from the wide-field camera and the other instruments to one of the relay satellites for transmission to Earth. In this way, the CCD turns light that has traveled millions of light years into electrical charges that tell astronomers on Earth how much light has reached each of the pixels within a given interval of time. We can use this information to reconstruct an image of the object under observation that is far sharper than anything obtainable on Earth.

Because only part of the telescope's incoming beam of light reaches the wide-field camera, and because the CCD detector spans "only" one inch on a side, the wide-field camera captures a field of view that has only one-tenth of the total diameter of the Space Telescope's field of view, 2.7 minutes of arc for the camera out of a total of 28 minutes of arc for the entire telescope. This still represents quite a wide field of view by astronomical reckoning, so the camera is indeed a "wide-field" instrument.

Furthermore, there is another way that astronomers use the wide-field camera, a way that explains the second half of the camera's full name. The wide-field/planetary camera's complex name stems from the fact that this amazing instrument can operate in either of two complementary modes. Besides the wide-field mode of the camera, it can operate at what amounts to a higher magnification, using an additional reflection that lengthens the path that the light travels to form an image. This increases the telescope's focal length by a factor of about 2.5, producing a larger image. As a result, in its planetary mode, the camera has a smaller field of view—slightly greater than one minute of arc across—but can see finer details than in the wide-field mode.

The planets Mars, Jupiter, and Saturn each have angular sizes of about half a minute (thirty seconds) of arc. Thus, in its planetary mode, the wide-field camera will be well matched to the planets' sizes, and should provide a continual stream of images whose clarity nearly equals those from the best planetary spacecraft flybys. Each image can be obtained within a few seconds of time, which should enable astronomers to obtain excellent pictures of fleeting features such as the fast-changing cloud patterns of Jupiter's atmosphere. This is particularly impressive when you recall that obtaining such sharp pictures previously required sending an expensive spacecraft hundreds of millions of miles into space.

The camera's planetary mode will be used not only to observe planets, but also to obtain the highest-resolution photographs of distant galaxies. In this mode, which takes the fullest possible advantage of the CCD detector's pixels and of the Space Telescope's near-perfect mirror, the wide-field camera should become famous—as famous as any astronomical camera ever has been.

The CCD detectors that form the camera's heart deserve attention for an additional fantastic ability. They can detect not only the visible light that strikes them, but also light whose wavelength places it deep in the ultraviolet portion of the spectrum. Furthermore, the CCDs can even detect light on the other side of the visible region, light in the near-infrared spectral region, that is, with wavelengths longer than those of visible light. Most CCDs are sensitive only in the visible and infrared spectral regions, which is fine for Earth-based observatories, where ultraviolet never penetrates.

To take advantage of the fact that the Space Telescope receives ultraviolet radiation as well as visible light, engineers discovered that by coating CCD chips with coronene—an organic phosphor, similar to the compounds that glow in the dark on fireflies—they could make the CCDs convert ultraviolet radiation into visible light, which the sensors can indeed detect. As a result, the Space Telescope has the finest CCDs ever made, with an unrivaled range in wavelength.

Suppose that the wide-field camera is studying a galaxy that we know to contain hot stars, which radiate primarily ultraviolet, as well as cool stars, which radiate long-wavelength visible light and infrared. In the camera's normal operation, all of these types of radiation would be converted into a single type of electrical signal. Thus, when we study the image of the galaxy, we would have no way to distinguish the galaxy's hot from its cool stars. But the wide-field camera is provided with forty-eight different light filters, each of which can be inserted into the light path as light enters the camera. Each filter removes from the beam of incoming light all the radiation *except* that within a narrowly set band of wavelengths centered on one particular reference wavelength. Thus, the camera can obtain several different images of a galaxy that correspond to different filters in the light path. Comparison of these images will reveal the locations of the high-temperature and the low-temperature stars, because the radiation from the two types of stars tends to fall in different wavelength bands.

THE WIDE-FIELD CAMERA AND SERENDIPITY

Astronomers, even more than most scientists, are fond of the concept of serendipity—the sort of thing that occurs when you drill for water and strike oil. Since the wide-field camera will operate at all times, even when astronomers are not directing it to look at any object in particular, the data sent from the camera back to Earth and stored for future reference may prove marvelously productive in a serendipitous sense.

Suppose, for example, that we are studying a bright star with the Space Telescope's high-resolution spectrograph, in an attempt to determine what chemical elements are found in the star. At the same time, simply because it is there, the wide-field camera will be recording all of the stars and galaxies that happen to lie in nearly the same direction as the bright star. Anyone who examines these data would, at the very least, be able to study a random portion of the sky with a sharpness never before available. An astronomer on the ground might choose to do this in the hopes of finding a previously unsuspected object. Since the data reach Earth in the form of electromagnetic impulses, it is a simple matter to store the data for later computer analysis to see if anything odd appears in the wide-field camera's images.

After several years have passed, the camera will have covered a few square degrees—about 0.01 percent of the total sky—and astronomers who want to find out generally what's out there can refer to the collection of these images to do so. Amateur astronomers who choose to perform the arduous and time-consuming task of examining these images may well discover new minor planets, new quasars, or even types of objects now unknown to us. The archive of data from the wide-field camera could conceivably prove—for reasons now unsuspected—the greatest treasure obtained with the Space Telescope.

THE FAINT-OBJECT CAMERA

In addition to the wide-field camera, the Space Telescope contains a second camera, which provides a set of different capabilities as well as a backup to the wide-field camera. The entrance to this second camera, called the faint-object camera, can also receive the light that passes through the hole in the primary mirror. The faint-object camera achieves a higher resolving power—a still sharper image—than the wide-field camera, by using a still longer focal length.

In fact, the faint-object camera consists of two separate cameras, each of which deflects light along its own separate path with mirrors that project into the beam of light that enters the system. The faint-object camera uses detectors that are not CCDs but image-intensifying devices based on the same principle that was used in early models of television cameras. Unlike the CCDs in the wide-field camera, an image-intensifying device counts individual photons. This detector can therefore achieve a sensitivity of detection even greater than that of the wide-field camera. In fact, the image-intensifying devices in the faint-

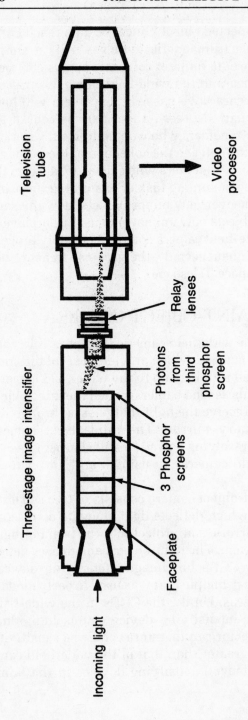

Figure 32: A schematic of the image intensifier, showing how each photon on the faceplate generates many photons that can be recorded by the television tube. *Drawing by Marjorie Baird Garlin*

object camera come close to achieving the highest sensitivity allowed by the laws of physics.

The faint-object camera's image intensifiers are each something like a television screen. A television screen glows when electrons guided by magnetic fields strike the phosphorescent material on the back of the screen. In the faint-object camera, each photon that strikes the faceplate of the detector produces an electron. The image intensifier adds millions of such electrons to that single electron, and then sends the entire bunch of electrons to a sort of television screen that can be read by on-board electronic equipment (Figure 32).

In a CCD detector, each pixel requires not one but about ten photons in order to produce a detectable signal, but the faint-object camera can detect individual photons. In simplified terms, the CCD can count photons only ten by ten, but the faint-object camera's image intensifier can count them one by one. As its name implies, the faint-object camera will be used primarily to observe the faintest objects, as well as those for which the greatest possible sharpness of image is required. The far-distant galaxies that lie at the fringes of the visible universe, ten to twenty billion light years from Earth, are prime targets of observation for the faint-object camera. Astronomers hope that these observations will reveal much about how galaxies formed and evolved, billions of years ago, at the time when their light began its long journey toward the Space Telescope.

Other objects detectable only with the faint-object camera are the white dwarf stars that are believed to exist in globular star clusters. Such globular clusters each contain many hundred thousand stars, all born at the same place and at nearly the same time (Figure 33). The stars in such a cluster burn themselves out at different rates, with the most massive stars the first to die. We therefore expect that within a globular cluster, a significant fraction of all the stars—perhaps 1 to 5 percent—have already become white dwarfs, the dying embers of once-active stars, now denuded of their outer layers. Not a single one of these white

Figure 33: Omega Centauri, a typical globular cluster of stars. *National Optical Astronomy Observatories*

dwarfs can be seen with an Earth-based telescope, for they shine too dimly to be visible amidst the sea of the brighter, still active cluster members. But the faint-object camera on the Space Telescope should be able to detect most if not all of them. This detection will allow us to verify our hypotheses about the rate at which stars of a given mass evolve into white dwarfs, and to determine what fraction of all stars have already become white dwarfs, as our sun someday will.

SPECTROSCOPY: RESOLVING LIGHT INTO ITS COMPONENT COLORS

To appreciate the remaining two instruments aboard the Space Telescope, we must pause to deepen our understanding of the importance of *spectroscopy*, the analysis of light by its various colors. A large part of an astronomer's work—at least for those who observe the stars rather than theorize about them—consists of separating light into its components, organized by wavelength. This spectroscopic analysis, whose name refers to the spectrum of different wavelengths, has an importance far greater than bean-counting classification. The study of light wavelength by wavelength reveals many of nature's deepest mysteries. In particular, such studies show us what the visible universe is made of.

Deep within every star, thermonuclear fusion reactions among nuclei release the energy of trillions of hydrogen bombs every second, as protons (hydrogen nuclei) fuse into helium nuclei. The energy released by these fusion reactions heats the star's center, giving the particles there extra energy of motion. These fast-moving particles collide with other particles somewhat more distant from the center, and these particles in turn collide with other particles still more distant. Eventually the gas in the entire star, heated by the nuclear fusion at its center, attains temperatures that range from ten or twenty million degrees at the center to several thousand degrees at the star's surface.

The heat within the star maintains the entire star as a glowing ball of gas. If we could descend into such a fiery hell, we would find particles—protons, helium nuclei, and electrons—dashing to and fro at enormous speeds. We would also find that, simply because it is so hot, the gas shines with electromagnetic radiation—light of all wavelengths. In fact, at a temperature above absolute zero, every object emits radiation. The cooler objects—our own bodies, for example—emit longer-wavelength radiation, primarily infrared and radio waves. The hotter objects—stars, for instance—emit shorter-wavelength light, primarily ultraviolet and visible light, along with infrared radiation. Extremely hot or violent objects emit x-rays and gamma rays.

The center of a star, which has a temperature of millions of degrees, ranks among the hottest objects in the universe, and produces mainly gamma and x radiation, the radiation with the highest energies and shortest wavelengths. If a star had only a center, this radiation would escape directly into space, bathing our Earth not in the sun's ultraviolet and visible light, but instead in much more energetic and dangerous radiation of shorter wavelengths. However, the center of every star, including our sun, is surrounded by hundreds of thousands of miles of gas. This gas traps the gamma and x radiation, saving us from destruction. Because the short-wavelength radiation is blocked by the gas, it gives up its energy to the gas and heats it. Since the temperature outside the star's center does not rise quite as high as it does at the very center, the intermediate layers of the star are primarily filled with ultraviolet radiation. This radiation in turn is blocked by gas lying still farther from the star's center, and likewise heats the gas surrounding it (Figure 34). Finally, the gas near the star's surface has temperatures at which primarily ultraviolet and visible light are emitted. What we call a star's surface is the region from which radiation has a good chance of escaping into

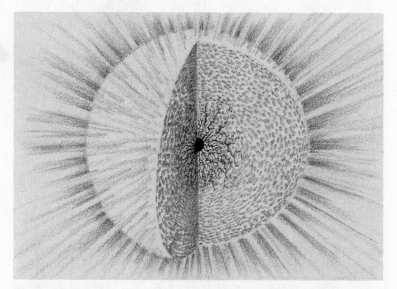

Figure 34: A schematic diagram of the interior of a star. The surface glows at a temperature in the range of 4,000 to 70,000 degrees, while the center may be at 10 to 20 million degrees. In between, the temperature varies smoothly between the central and surface temperatures.

space instead of being completely blocked by the gas above it.

When you think of a star, think of a vast cauldron of hot gas, fantastically hot at the center, merely enormously hot at the outside. The energy released in thermonuclear fusion at the center diffuses outward through countless processes of trapping and re-emission of electromagnetic radiation. Finally, when the radiation has penetrated enough of the star's material that relatively little gas lies beyond, it has a chance to pass in a straight line outward into space with no further blockage. When we see the sun's surface, we see the lowest layer from which radiation escapes directly. Further in than this we cannot see, simply because there the gas is thick enough to block direct escape of the radiation.

But a strange thing happens to this electromagnetic radiation in its last dash outward. Because the final passage occurs through regions that have temperatures of "only" three thousand to fifty thousand degrees on the absolute scale (four thousand to seventy thousand degrees Fahrenheit), this escaping radiation passes through parts of the star—its outermost layers—that are cool enough for atoms, ions, and molecules to exist. At temperatures of half a million degrees or more, no atoms or molecules can exist; instead, high-energy collisions strip all the electrons loose, leaving only the nuclei, with the electrons sailing freely among them. Such completely ionized gas can block radiation—as we have just described—but it does so relatively inefficiently compared to a gas made of atoms or of ions that have not lost all their electrons. Furthermore, completely ionized gas blocks radiation of all wavelengths about equally well.

Atoms, ions, and molecules block radiation quite differently than a completely ionized gas does, blocking only radiation having definite, particular wavelengths. Each type of atom, ion, or molecule—be it a hydrogen, helium, carbon, nitrogen, or oxygen atom or ion or a carbon monoxide, water vapor, or sulfuric acid molecule—blocks a particular set of wavelengths and no others. The unblocked types of radiation pass among the atoms, ions, and molecules with ease, unaffected. But the radiation at the particular wavelengths that the atom, ion, or molecule does block is removed with great efficiency from the beam of light.

Therefore, when you study light from a star, what you see is light of nearly all wavelengths, ultimately tracing its energy to nuclear fusion at the star's center. But some of the light is missing—the light removed during the last dash of the radiation into space by the types of atoms, ions, and molecules that exist near the star's surface.

The fact that each particular type of atom, ion, or molecule removes a particular set of wavelengths places a powerful tool of physics at astronomers' disposal. If astronomers can recognize just which set of wavelengths has been removed from a beam of light, they can tell which types of atoms, ions, or molecules have performed the removal. In other words, despite the thousands of light-years (or even millions in the case of galaxies) that light may have traveled, astronomers can detect which types of atoms or molecules exist in the outer layers of stars. We can use the spectroscopic fingerprint of wavelength removal to detect what the stars and galaxies are made of (Figure 35).

Spectroscopy refers to the analysis of objects by means of the wavelengths of the radiation that the objects remove from a beam that originally contains all wavelengths. Spectroscopy requires a tool (called a spectrograph) that divides radiation into its constituent wavelengths and thus allows scientists to see which wavelengths have been removed. (In some other situations, an extra amount of certain wavelengths of radiation has been *added*.) Throughout the past century, astronomers have built progressively more accurate and sensitive spectrographs and have mounted them on better and better telescopes, in their continuing effort to analyze the composition of the celestial objects that they observe. This effort will reach a peak aboard the Space Telescope.

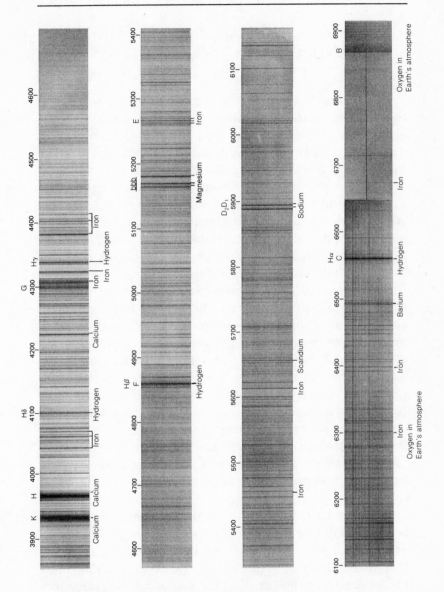

Figure 35: The sun's spectrum in visible light shows a host of dark absorption lines at different wavelengths (denoted in ten-billionths of a meter). The names beneath some of the lines identify the chemical elements that produce them. *Mount Wilson and Las Campanas Observatories*

THE FAINT-OBJECT SPECTROGRAPH AND THE HIGH-RESOLUTION SPECTROGRAPH

The Space Telescope carries not one but two spectrographs. One of them, the faint-object spectrograph, is designed to measure spectra of the faintest possible objects in a relatively crude way. Spectrographs differ in the discrimination with which they can distinguish between neighboring wavelengths. More discrimination gives astronomers the chance to see subtle differences between neighboring wavelengths, subtleties that will pass undetected if the spectrograph observes each bunch of neighboring wavelengths as a single unit. In theory, wavelengths could be measured with near-infinite accuracy. In practice, the more finely we seek to measure each exact wavelength, the more time we must spend to obtain the object's total spectrum. Hence the faint-object spectrograph sacrifices some fineness in its measurement of the spectrum in order to obtain at least some spectral results for faint objects.

In contrast to the faint-object spectrograph, the high-resolution spectrograph, officially called the Goddard High-Resolution Spectrograph, can be used to measure the spectrum only of relatively bright objects. That is because the high-resolution spectrograph divides the light so finely into its constituent wavelengths that each individual wavelength band contains only a relatively small amount of light.

The faint-object and high-resolution spectrographs are both mounted behind the primary mirror of the Space Telescope; each receives all the light that is reflected by that mirror in the direction of that spectrograph. The faint-object spectrograph can discriminate among wavelengths with a precision of one part in a thousand; that is, any two neighboring wavelengths for which the spectrograph measures the brightness of light differ by 0.1 percent in their wavelengths. Wavelengths closer together than this will be detected as a single wavelength. The highest resolution available with the high-resolution spectrograph is a fineness in wavelength discrimination that is a hundred times better—one part in 100,000.

This spectral resolution exceeds that of any other spectrograph in space, such as the spectrograph on the *International Ultraviolet Explorer* satellite. Hence the high-resolution spectrograph will break new ground in our spectroscopic studies of the cosmos, revealing details of chemical composition in stars and galaxies that may, in their turn, shed new light on how stars and galaxies began to form and how they have evolved through billions of years.

In both the faint-object and high-resolution spectrographs, light is split into its various wavelengths by a diffraction grating, a mirror on which thousands of closely spaced parallel lines have been ruled (Figure 36). Because of the wave properties of light, light reflecting from the mirror will be broken up by the grating, so that each wavelength is reflected in a slightly different manner. Thus, the reflection tends to separate the radiation of each individual wavelength. The separate wavelengths strike the detectors at slightly different places, so they can each

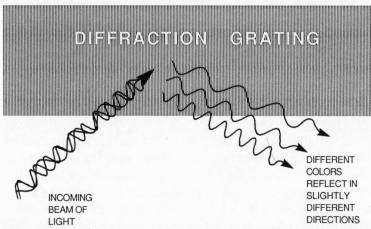

DIFFRACTION GRATING

INCOMING
BEAM OF
LIGHT

DIFFERENT
COLORS
REFLECT IN
SLIGHTLY
DIFFERENT
DIRECTIONS

Figure 36: A schematic of a grating spectroscope shows the finely ruled parallel lines that reflect light of different colors in different directions. *Drawing by Marjorie Baird Garlin*

be measured separately from all the other wavelengths. In both of the spectrographs on the Space Telescope, the radiation detectors are coated surfaces that emit electrons when they are struck by visible light or ultraviolet; these electrons can be detected and measured, revealing how much radiation of each wavelength has impinged on the detector.

The faint-object spectrograph can detect radiation with wavelengths over a range that covers the ultraviolet and visible-light portions of the electromagnetic spectrum, but the high-resolution spectrograph detects only ultraviolet photons. Why this difference? The high-resolution spectrograph performs a task similar to what spectrographs do when mounted on ground-based telescopes, but these telescopes can detect and analyze only visible-light photons. Since the high-resolution spectrograph has no better resolution than the best Earth-based spectrographs, no reason exists to spend the Space Telescope's time repeating what we can do from Earth-based observatories. The high-resolution spectrograph is therefore specialized to work only in the ultraviolet. On the other hand, the Space Telescope will detect objects far fainter than those seen from the ground in visible light. Hence it makes sense to use the faint-object spectrograph to carry out spectroscopy not only with ultraviolet radiation but also with visible light from objects too faint to observe from the ground.

The faint-object spectrograph, in some respects, represents the ultimate thrust of the Space Telescope; objects too faint to be seen at all from ground-based observatories will not only be detected, but will also be analyzed spectroscopically, in both visible light and ultraviolet, with a fineness in wavelength of one part in a thousand. One example of such an invisible object would be a quasar (quasi-stellar object), a pointlike source of immense amounts of energy, ten or fifteen billion light-years away. Many quasars have already been detected, but many other, fainter quasars are believed to exist, awaiting discovery by the Space Telescope. If this occurs, we shall not only see these as-yet-undetected quasars, but shall also be able to determine their chemical compositions through spectroscopic analysis with the faint-object spectrograph. In other words, we can look backward in time through ten or fifteen billion years, to see how the chemical composition of quasars then differed from the composition of the much closer quasars, "only" a few billion light-years away, that we have already analyzed.

The high-resolution spectrograph will also allow scientists to learn more about the interstellar gas that spreads diffusely between the stars, both within our own Milky Way Galaxy and (in much lower density) between galaxies. Many of the types of atoms and ions that exist in this interstellar gas do not block visible light at all. Such types of atoms and ions can be detected and studied only by observing the types of ultraviolet radiation that they *do* block. Hence we need a spectrograph with the capacity to perform ultraviolet studies, and we need one with high resolution in order to determine just which types of atoms exist in the diffuse gas that lies among the stars.

PHOTOMETRY: MEASURING THE BRIGHTNESS OF OBJECTS

In addition to spectroscopic analysis, astronomers usually want to measure as precisely as possible the apparent brightness of the objects they observe. The reason is that since the apparent brightness of any object decreases in proportion to the square of the distance between the object and the observer, more distant objects generally tend to appear fainter than those closer to the observer. The application of this principle plays an important role in astronomers' research, because it allows them to reach significant conclusions about how stars and galaxies are distributed through space.

But photometry, the measurement of apparent brightness, has an aspect that goes beyond measuring an object's apparent brightness. By inserting filters that allow only certain wavelengths to pass through, astronomers can measure how bright an object is in yellow light, how bright in red, how bright in green. By comparing these measurements, they can perform a rough-and-ready spectroscopic analysis, one that proceeds hundreds of times faster than "true" spectroscopy can. When we consider the prospect of examining, say, one thousand stars, we realize that it is not realistic to measure the full spectrum of each star. But to measure each of the stars' apparent brightnesses in three or four colors—which amounts to compressing the spectrum into not one thousand separate bands but just three or four—takes a remarkably short time, so short that even one thousand stars can be dealt with in this summary fashion during a few dozen hours of observation.

This fact is so because the high-speed photometer (HSP) aboard the Space Telescope can measure the apparent brightness of a relatively bright star in one hundred-thousandth of a second! Furthermore, the high-speed photometer measures not in three or four colors, but in hundreds of different colors. And it does this without a single moving part! Instead, the light that enters the high-speed photometer, which is mounted behind the primary mirror along with the faint-object camera and the high-resolution spectrograph, falls on many different openings within the photometer system. Each opening has a different combination of filters; by measuring the amount of light that passes through each opening, the high-speed photometer can perform a not-so-crude spectroscopic analysis in far less time than the blink of an eye.

The photometer's ability to work rapidly proves valuable in yet another way. Some astronomical sources—radio pulsars, in particular—are known to vary in brightness many times, sometimes hundreds of times, each second. The record rate of variation, for the pulsar 1937 + 214, reaches 642 times per second. There is a report—unconfirmed as of this writing—that a light-emitting pulsar associated with the remains of Supernova 1987A emits nearly 2,000 pulses of light per second. The high-speed photometer will prove highly useful in studying such objects, as well as others that may vary in brightness even more rapidly, including—according to some theories—black holes, or more precisely the regions immediately surrounding black holes. Hence the high-speed photometer may provide a roundabout but important way to reveal the existence of black holes, which may prove to be abundant in a galaxy such as our own.

OTHER BENEFITS OF THE FINE GUIDANCE SYSTEM

The fine guidance sensors, described in Chapter 4 as part of the means of keeping the Space Telescope properly pointed, offer yet an additional way to make astronomical observations. After all, the pick-off mirrors used in the fine guidance sensors always take part of the light so carefully gathered by the ninety-four-inch mirror, and it would be a shame to spend this light entirely on the task of guiding the telescope. Instead, since the whole point of the fine guidance sensors is to locate objects with incredible accuracy, we can use our observations to establish, with this kind of accuracy, the positions of the objects we observe. In other words, the act of guiding can yield the positions of stars on the sky more accurately than we can now measure them, because the atmospheric blurring ruins our best measurements from ground-based observatories. In practice, two of the three fine guidance sensors do the guiding, while the third measures star positions accurately.

Thus, in addition to its five main instruments—the wide-field camera, the faint-object camera, the high-resolution spectrograph, the faint-object spectrograph, and the high-speed photometer—the Space Telescope can employ its fine guidance sensors as a sixth instrument to study the cosmos. Quite a complement of equipment! But no more than a $2 billion telescope deserves.

6

THE SCIENTIFIC
ENTERPRISE

SEEN FROM OUTSIDE the scientific community, the Space
Telescope is a mighty, automated instrument in space.
However, seen from the perspective of science as an institution, the Space Telescope is not just the instrument itself
but a *system* for doing science—a system that includes not
only the instrument, but an institute to run it, a federal
agency (NASA) that is responsible for keeping it going, the
European Space Agency that has a 15 percent interest in it,
and an international group of astronomers who use it. The
outsider and insider perspective are both necessary to a
true view of the Space Telescope.

THE FLOW OF DATA

The true extent of the Space Telescope system appears
clearly when we consider how data from the Space Telescope flow from the satellite to the ground, and then diffuse
outward to the astronomical community and the general
public. As we described in Chapter 4, the Space Telescope
stores its data temporarily and then passes the data to one
of the relay satellites that ride high above the Earth in
synchronous orbit. The relay satellites send their data back
to Earth—but to where? The answer is interesting: the data
flow to *two* locations, one governmental and one not so
governmental.

The Space Telescope data go to the Goddard Space Flight Center and to the Space Telescope Science Institute. Goddard, as part of NASA, bears the responsibility for the safe and effective operation of the Space Telescope, which was paid for by American and European taxpayers. The institute, operated for NASA under contract with the Associated Universities for Research in Astronomy (AURA) is a hybrid creature. The institute was created by the astronomical community and AURA with NASA support, and is designed to provide an effective liaison between NASA, ESA, and the international community of astronomers. The tension between the two aspects of the Space Telescope—as the product of NASA and ESA on the one hand, and as a tremendous tool for advancing our scientific knowledge on the other—works itself out in the discussions between Goddard and the institute about just who should control precisely which commands going to the Space Telescope and which data arriving from the Space Telescope.

In a rough and ready way, protracted discussions between the astronomers and the space agency have yielded the following solution: Goddard has charge of keeping the satellite functioning, and the institute will direct its scientific use. In the vagueness of these words lies great opportunities for argument. But they will serve as a general principle, to be matured into operational reality through experience.

THE PROPOSAL PROCESS

How does a scientist make the Space Telescope do his or her bidding? With this question, we enter the world of scientific proposals, a strange arena of paperwork, persuasion, and power. All large scientific facilities—the particle accelerators, the advanced biological laboratories, the ground-based telescopes—receive far more demands for their use than they can actually fulfill. In the old days, before the government played an important role, the inherent conflict of "too many pigs for the teats" (to use Abraham Lincoln's phrase) was solved by brute force. The director of the observatory, for example, decided whose observing projects were approved and whose rejected. This system had the virtue of simplicity, although sometimes the staff rose in revolt and achieved the dismissal of the director for his perceived prejudice, demonstrated in his choice of projects, among other managerial failures. Such revolts led to a change of directors, but not to a change of system.

After long years of experience, a different system has finally evolved, an intriguing mix of democracy and elitism, which now governs the assignment of time on the largest pieces of scientific equipment. Under this system, all scientists are theoretically equal; they are all entitled to submit proposals to those who run the machines, explaining what they will do if they have a chance to use the equipment. A time allocation committee reviews each proposal and, in effect, grades it. The A proposals get all the time requested, the B proposals some of the requested time, and the Cs a polite rejection. The membership of the time allocation committee therefore determines the present pattern, and thus the future, of scientific research with a given instrument.

And who are the committee members? Here lies the secret of modern government. Well-known scientists typically dominate such powerful bodies. Although the scientists on the committee could simply allot the observing time to their friends and to no one else, this rarely happens. For one thing, the committee members are eager above all to see their field of research advance, so they vote on the basis of whether a given proposal is likely to advance science or not. And to eliminate the most obvious source of bias in the voting, members of the committee who are based at the institution from which a proposal arrives are not permitted to vote when that particular proposal comes up.

THE TIME ALLOCATION COMMITTEE OF THE SPACE TELESCOPE

So it goes with the Space Telescope, but with a slight wrinkle. The Space Telescope Science Institute has created a Time Allocation Committee (TAC) to decide which scientific proposals will receive observing time on the Space Telescope. In total, the TAC has at its disposal 3,000 hours per year (roughly eight hours per twenty-four-hour day); the remainder of the Space Telescope's time will be devoted to nonobserving tasks or has been allocated in advance. That is the basic plan; the wrinkle is that the director of the institute, Riccardo Giacconi, has the right to overrule the TAC. But any good director knows that you don't do this often, or you will find yourself in need of a new committee, and eventually of a new job.

How is it that only 3,000 hours per year (out of a total of nearly 9,000) are available to be allotted by the TAC? To understand this, visualize how the motion of the Space Telescope in its orbit affects its ability to collect data. Consider, as an analogy, the limitations on ground-based observatories. Ground-based optical observatories can work only at night, since during the day the sunlight reflected by our atmosphere makes virtually all astronomical objects but the sun invisible. But where the Space Telescope orbits, high above nearly all of the Earth's atmosphere, this problem disappears, because there is no atmosphere there to reflect sunlight. Astronauts have commented on the strangeness of a perfectly black sky, with the stars easily visible even when the sun is above the horizon.

Unfortunately, the absence of atmospheric reflection at orbital altitudes does not mean that the Space Telescope can be used for twenty-four hours each day. The limit arises from the Earth. To appreciate the effect of the Earth on scheduling of observations with the Space Telescope, pretend that you are an astronaut circling the Earth every ninety-six minutes in an orbit like that of the Space Telescope. Taking off in the Space Shuttle from Cape Canaveral at dawn, say, you see the sun just rising above the eastern horizon. Had you remained on the ground, the sun would have appeared to rise slowly, moving about fifteen degrees across the sky in an hour, owing to the Earth's eastward rotation. But as you travel eastward in the Space Shuttle at seventeen thousand miles per hour, you see the sun rising at nearly four degrees per *minute*; almost before you realize what is going on, it is already "noon" from your vantage point, and then the sun sets in the west only forty-eight minutes after you took off! The Earth is plunged into darkness beneath you for an "orbital night" that lasts only forty-eight minutes. Then once again you see the sun rising, as you did at takeoff.

Thus, the entire day-night cycle lasts only ninety-six minutes, rather than the twenty-four hours familiar to us on Earth. The reason for this rush through day and night is that any object in a low Earth orbit must move in that orbit fifteen times more rapidly than the Earth rotates beneath it. If the orbiting object traveled any slower, it would fall to Earth, unable to resist the Earth's gravitational pull.

Now suppose that while aboard the Space Shuttle, you attempt to perform some astronomical observations with your binoculars. Soon after takeoff, delighted by the fact that the sky is dark (although the sun is still up), you train your binoculars on the Orion Nebula, a cloud of interstellar gas about 1,400 light-years distant, heated to a glowing temperature by hot stars within it (Figure 37). It happens that at the time of year of your flight, the Orion Nebula is ninety degrees west of the sun, so the nebula is nearly overhead as you take off. But, like everything else in the sky, the Orion Nebula appears to move westward at nearly four degrees per minute and sets in the west only twenty-four minutes after you first observe it. Obviously, no more observations of the nebula are possible until it rises again, forty-eight minutes later. This rhythm of forty-eight above the horizon and forty-eight minutes below will dominate the Space Telescope observations, so that continuous ob-

Figure 37: The Orion Nebula, a region where hot stars make nearby gases so hot that they glow. The superposed dark lanes are clouds of interstellar dust. *Lick Observatory*

servations of more than forty-eight minutes are simply not possible. From another perspective, every forty-eight minutes, you find yourself on the "wrong" side of the Earth to observe a particular object.

Actually, this description is accurate only for objects that happen to lie in the plane of the orbit. Objects in a direction perpendicular to that plane are visible nearly all of the time, while objects at intermediate positions with respect to the plane of the orbit are visible for more than forty-eight minutes but less than ninety-six minutes per orbital period.

In principle, you can swing the telescope (that is, rotate it around one or more axes) to target a new object as soon as the previous one sets in the west. Because you want to catch any object while it is rising, in order to obtain the maximum time to observe it, the telescope must be swung from the western to the eastern horizon. This requires about thirty minutes, a substantial fraction of the time that an object is above the horizon. This is already bad enough, but actually the Space Telescope suffers much greater losses of observing time because it often will be used to make exposures more than one orbital period long on any given target. Given the inefficiencies in swinging the telescope around and acquiring a new target, the telescope will usually not be moved while the target is being obscured by the Earth. Instead, one simply waits until the target rises again, and because the telescope has not moved, it will already be trained on the target, ready to continue the exposure.

For those targets to be observed for more than one orbit, therefore, much of the time is wasted waiting for the Earth to get out of the way. When additional time is added for swinging the telescope and acquiring new targets, it proves quite difficult to achieve more than about 30 percent efficiency with the telescope. Thus, the best estimates are that the Space Telescope will obtain data on one target or another for only about 30 percent of the total elapsed time, or an average of under eight hours per day.

By coincidence, this is about the length of time available for astronomical observations with a ground-based telescope, mostly because of the brightness of the sky during the daytime and during the twilight before sunrise and after sunset. We will not put a telescope into space to gain extra time—only extra clarity.

MAKING A PROPOSAL TO THE TIME ALLOCATION COMMITTEE

To receive approval from the Time Allocation Committee (TAC), each scientist or group of scientists must submit to the Space Telescope Science Institute a proposal explaining what they want to observe, why, with what instruments, and for how long. To understand how this approval process works, let us look at the proposal submitted by Professor Robert Kirshner of the Harvard-Smithsonian Center for Astrophysics and his colleagues in response to the institute's first call for proposals in August 1988. The proposal itself appears as Figure 38, where we show the first of dozens of pages.

PROPOSAL FOR OBSERVATIONS
HUBBLE SPACE TELESCOPE
COVER PAGE
(see instructions on back)

1. Proposal Title:
The Young Remnants of Massive Supernovae

2. Scientific Category (select one):	3. Proposal Category:	4. Proposal Type (if applicable):	5. Proposal References (if applicable):
☐ Solar System	☒ GO Time	☐ Large Project (☐ Key Project)	
☒ Interstellar Medium	☐ GTO Time		☐ Continuation of HST
☐ Stellar Astrophysics	(Team: _____)	☐ Long Term (_____ years)	Program Number: _____
☐ Stellar Populations	☐ Archival Research		
☐ Galaxies & Clusters	☐ Other _____	☐ Target of Opportunity	☐ Remote Submission
☐ Quasars & AGN			File ID: _____

6. Principal Investigator Name:	Institution:	Country:
Robert P. Kirshner	Harvard College Observatory	USA

7. Abstract (please confine to this space):
The remnants of recent supernovae provide the best opportunity to probe the evolution of massive stars and the synthesis of heavy elements. Nine young remnants with fast-moving, undiluted debris are known, including two we have discovered in 1988: 4 in our Galaxy, 3 in the Magellanic Clouds, one in NGC 4449, and one in M83. We have obtained ground-based spectra and images of the young remnants, employed IUE to its limits, and developed new data analysis tools to study them. The results provide valuable insights into the ages, composition, and kinematics of young remnants, but are incomplete in tantalizing ways that HST can resolve. While we are confident these remnants result from the violent destruction of massive stars after advanced nuclear burning, essential features of the explosion physics, the excitation of the debris, the chemical composition of the ejecta, and the age, distance, and kinematics still elude our grasp. HST observations, both images and spectra, will allow us to isolate the chemical inhomogeneities in the debris and obtain UV spectra of individual filaments. The images will allow an unprecedented probe of the excitation mechanism, and will provide 10 times the angular resolution for proper motion studies to determine ages for the most distant objects.

8. Scientific Key Words:
Abundances, Nucleosynthesis, Supernova Remnants

9. Total Exposure Time Requested:	41 hours (Primary)	10. Number of Targets:	9 (Primary)
	_____ hours (Parallel)		_____ (Parallel)

11. Scientific Instrument(s) Requested: ☒ WF/PC ☒ FOC ☒ FOS ☐ HRS ☐ HSP ☐ FGS

12. Special Scheduling Requests: ☐ Real-Time Observations ☐ Time-Critical Observations

13. Spacecraft Time and Program Efficiency:	16. Authorizing Institutional Official:
	Name:
Estimated Spacecraft Time: 64 hours	Title:
Program Efficiency: 64 %	Institution:
14. Funding Request (U.S. scientists only):	Address:
Funds (total): $ 343,390	
Length of Funding Period: 12 months	
Starting Date (month/year): Jan. 1991	City: State/Province:
15. Principal Investigator Title:	Postal/Zip Code: Country:
Professor of Astronomy	Telephone: Telex:
Signature: _Robert Kirshner_ Date: 26 Sept 88	Signature: _____ Date: _____

Figure 38: Prof. Robert Kirshner's proposal to use the Space Telescope to study the young remnants of supernova explosions. *Prof. Robert Kirshner, Harvard University*

Kirshner proposes to observe supernovas (exploding stars) in other galaxies; he stresses the use of the Space Telescope's ability to make observations with high spatial resolution and of ultraviolet radiation. Seven scientists in addition to Kirshner are involved in the proposal. This reflects the fact that doing astronomy with the Space Telescope is a highly competitive business, so it is necessary to demonstrate to the TAC that the scientific team has competence at several different levels.

First and foremost, the members of the TAC will ask whether the project is scientifically worthwhile. Even presuming that all will go as projected (often not the case in science, as in any other activity), will the proposed observations produce new insights into astronomical problems of current interest? Of course, a considerable amount of scientific judgment and taste enter into the response to this question. Hence one or more members of the TAC must be experts in the field, who know what questions are important and why.

Next come the more instrumental aspects of the proposal. Should one use the faint-object camera or the wide-field/planetary camera to obtain images of the objects under study? What filters will be most effective, isolating the most interesting wavelengths of radiation from the objects under study? Will the greatest amount of information be obtained by using the highest spatial resolution available with the faint-object camera or by using the somewhat lower resolution, but higher sensitivity, available with the wide-field camera? Which wavelength resolution, on which of the two spectrographs, should be used to obtain the desired spectroscopic information—the faint-object spectrograph or the high-resolution spectrograph? To answer these questions effectively requires an astronomer who has worked intimately with the relevant instruments and has experience in making such choices on the basis of reasoned judgment. When the Space Telescope returns data from its observing session, the data will be in digital form, ready to be processed by computers. Does the proposed observing team include a person familiar with the best techniques for this data analysis?

The observing team should also include an astronomer who has a special interest in interpreting the data. For example, if supernovas are to be observed, a theorist who has been analyzing computer-generated models of supernova explosions might prove useful. As the data are analyzed, they can then be compared with recent theoretical models so as to draw the most useful conclusions from the observations. Do the observations contradict any current theory or make it less likely? Do they support some theory? How do they do so?

Most proposals to use the Space Telescope will be formulated by teams of scientists that include astronomers and astrophysicists with all the various backgrounds and strengths that are needed. Only in this way can a group of scientists hope to compete successfully for time on the Space Telescope. The effect is to encourage group efforts as opposed to individual projects—a logical result for allocating the use of a $2 billion instrument.

THE PROPOSAL'S CHANCES OF SUCCESS

If you as a scientist or your group of scientists submits a proposal, what are its chances of success? The proposal process is somewhat like an obstacle course on which your scientific group must surmount certain obstacles in order to succeed.

Obstacle 1. Prepare the proposal and be sure that it arrives at the Space Telescope Science Institute before the announced deadline for this round of proposals. This requirement may seem trivial but it is not. A key member of your proposal team may be out of the country, or a key piece of data may be missing. Add to these possibilities the well-known tendency for almost everyone to procrastinate, and you have a recipe for missing the deadline, a result that can sometimes be averted at the last moment by a race to type the proposal and take it to the Federal Express office before closing time on Friday afternoon. Some proposals will nevertheless arrive too late and so will not be considered until the next round of proposals, a full year later.

Obstacle 2. Pass a preliminary technical review by the institute staff. Some proposed targets for observation are so distant from any bright object on the sky that it will be impossible for the Space Telescope to target them by using the fine guidance system of the telescope (which requires not-too-faint stars close enough to use as guides to the target itself). Quick reference to a computer program listing all the known possible guide stars will reveal whether or not that possibility will prevent the observation. In addition, the proposed observations may rely on filters that should be in working order, according to manuals published previously by the institute, but that in fact are not currently working. Perhaps the sequence of observations proposed is inconsistent with the rules laid down to ensure the safety of the Space Telescope. For a host of such reasons, your proposal team may receive a brief message from the institute stating that a problem exists. In most cases, the team can work with the institute's staff to resolve such a problem, but in a few cases, the proposal will be held up at this stage.

Obstacle 3. Obtain a high rating for the proposal from the specific panel of the TAC that deals with the area of astronomy represented by your proposal. A common problem here will be that out of the some 5,000 astronomers engaged in active research worldwide, a few astronomers somewhere else will have teamed up to propose nearly the same project as yours. With different expertise represented by their team, they may have proposed some observation that you overlooked, for example, observation of a spectral feature that will yield a great deal of information. If they included this in their proposal, they may be rated higher by the panel for that reason. If you rate lower, your proposal may be rejected at this stage.

Obstacle 4. Even though your proposal is rated relatively high by the panel, it must also be rated high by the TAC as a whole. Here is one way that might not happen: If a flood of good proposals in your field of astronomy arrives at the TAC office, and fewer good proposals in other fields, the TAC could decide that because it is important to retain a balance among the different fields of astronomy, it will approve only the top 10 percent of the proposals in your field, but 15 percent in the other field. If your proposal falls just below the top 10 percent in your field, it could be rejected at this stage.

Obstacle 5. Even though your proposal is accepted by the TAC, it could nevertheless be turned down by the director of the institute. While outright rejection is extremely unlikely, the director might postpone a project if some emergency arises. For example, if you propose to observe well-known supernova remnants in other galaxies, and then just before your proposal is to be executed, a supernova explosion appears at a distance of only 160,000 light-years from the sun, your project will surely be postponed while the Space Telescope is directed to observe this once-in-a-lifetime phenomenon. (In fact, the bright supernova 1987A *did* appear in the Large Magellanic Cloud, only 160,000 light-years away, on February 23, 1987; had the Space Telescope been in orbit at the time, other programs would have been postponed so that the Space Telescope could be directed to observe the supernova in detail.)

Obstacle 6. Your proposal is approved, but before the actual observations are made, something goes wrong. Perhaps the instrument that you want to use experiences a minor failure, and the engineers must work for a few days to fix it, using commands from the ground. This possibility is inevitable in every space mission, and it could happen to you.

Suppose, however, that your proposal makes it safely through all the obstacles. Then, six months to a year after you first submitted your proposal, at a moment calculated by the institute to be the optimum one for your project (taking account of all the other demands on the telescope), the twelve-ton giant instrument will receive a stream of radio impulses sent via one of the relay satellites, telling it in which direction to point. The satellite's reaction wheels are engaged, and somewhere along its orbit, perhaps over Hawaii, the Space Telescope is directed to swing toward the direction that you requested. The telescope points rather inaccurately at first, but close enough to the desired position that the guide stars selected by the institute appear in the field of view. Working with the images of those stars, the fine guidance system corrects the telescope's aim until each of the guide stars is located just where it should be in the telescope's field of view.

If everything is working perfectly, the image of the object that you want to observe should fall at the entrance aperture to the instrument you have selected. If that does not happen, further adjustments can be made using the wide-field camera in its search mode, until the light from the object you proposed for observation finally enters the proper instrument. The Space Telescope's on-board computer has already turned on that instrument and has adjusted its internal parameters to the values that you chose in your proposal. Data begin to accumulate in the form of a stream of electronic signals, recorded on a magnetic tape much like the one in a VCR used to record television programs. When the Space Telescope next comes into view of the appropriate relay satellite, the data gathered by "your" instrument are read from the tape and transmitted to the satellite, which relays it to Goddard and the institute.

Every data packet must contain all the pertinent information about the Space Telescope itself, as well as your astronomical data about the object under study, for it is crucial that the flight controllers at Goddard know how the spacecraft is operating. Are the solar panels oriented properly to yield maximum electrical power from the sun? Are the voltages that operate the reaction wheels within a reasonable range of the designed values? Did the entrance slit of the spectrograph open at the rate at which it is supposed to open? These and hundreds of other facts are monitored constantly by electronic sensors on the spacecraft and are sent back to Earth along with the science data.

In a way, these "housekeeping reports" are even more important than the scientific data from any one project, because they tell the engineers at Goddard whether or not the spacecraft is healthy. If it is having problems, not only your project, but every project to follow is endangered until the problems are fixed.

The most dreadful thing that could happen to the Space Telescope is that the reaction wheels might impart too much spin to the spacecraft, so that it would begin to tumble end over end in space. Not only would this threaten the delicate mechanical systems on board, but the small radio dish antenna used to relay signals to and from the relay satellites might not be able to stay pointed toward a satellite, so that communication with Earth would be broken off. In that case, not only would the Space Telescope be in danger of self-destructing, but its engineers would be powerless to prevent this destruction, because they would have no way to communicate with the spacecraft. An automatic failsafe mode is therefore incorporated to prevent such an occurrence. One can only trust that it will work in all cases.

If all goes well, however, Goddard simply removes the housekeeping data for further analysis, and the scientific data pass over telephone lines to Baltimore, Maryland, home of the Space Telescope Science Institute. A light on the institute's computer will go on, and a stream of data that includes yours will pour in and be recorded on a magnetic disk.

DATA ARCHIVING AND REDUCTION

Here the Space Telescope Science Institute comes in to its own. Years of preparation have gone into devising computer programs that will turn your data into usable form. At this stage, the data are simply a string of off/on electronic impulses recorded on the computer disk. A typical observation with the wide-field camera may contain ten megabytes (ten million bytes) of data. Each byte consists of eight 'bits," or on/off impulses; a byte contains roughly the same amount of information as a letter of the alphabet. The ten megabytes of your data contain the equivalent of about ten million letters, or more than a thousand pages of information.

Because of the power of electronics, the disk that you see mounted on the computer's disk drive contains a thousand megabytes of data, of which ten megabytes are yours. How do you separate your data from all the rest? Fortunately, an institute staff member knows exactly what to do, because he or she was trained long ago to handle this situation.

The very first thing is to copy the information onto another magnetic disk (or else onto magnetic tape) for storage in an archive. No one wants to take the chance that the first—and only—disk with your data on it might be accidentally erased. There is also another, less obvious use for the archival copy. When your exclusive right to the data runs out (by agreement, this will happen one year after the data were obtained), the copy in the archive will become available to any qualified person for study. Here we see our system of government at its best. Since the taxpayers paid for the information acquired by the Space Telescope, it is theirs; they can come and look at it if they wish. However, if they don't know *how* to look at the data, they'll have to learn.

The operator then begins to analyze the information on the computer disk, using a program stored on another disk. In this process, the various observational projects carried out during the same time period as yours are sorted out, and the chief steps taken during each observation, such as changing filters, are identified in the data string. Another part of the program goes to work on the data itself. If your project aimed to obtain the ultraviolet spectrum of an object, the data arrive in the form of readouts from electronic detectors located at various positions along the area that the light reached, separated into its various wavelengths. Hence the data can be envisioned as an electronic readout for each of, say, 200 detectors.

To make the data useful, you need to know which detector corresponds to which wavelength in the ultraviolet spectrum and what the readout from each detector means in terms of the amount of light reaching it. Fortunately, the institute has this information available. Not only did the institute staff participate in calibrating each instrument before it was launched on the Space Telescope, but the same staff members have made it their business to keep track of the various engineering data on the operation of the Space Telescope sent back via the relay satellites, looking for any changes that may have occurred since the Space Telescope has been in orbit. So the operator calls up another disk on which calibration data are stored, and a computer program converts the detector readings to the intensities of light and the detector number to the wavelength of ultraviolet light. Then corrections are applied to any data that, based on experience with the instrument, may be faulty. The result is a graph and accompanying numbers that show the intensity at each wavelength for the object that you chose to study. This information is copied onto a magnetic tape that is mailed to you.

DATA ANALYSIS

About a year after proposing an observation, you receive a magnetic tape in the mail. The job of interpreting your data has only just begun. If you are part of a typical scientific or educational institution, a small but powerful computer or workstation, funded by NASA (or by ESA if you are in Europe), will be available to help you through the next steps in the analysis. Mounting the tape of the observations, you turn this computer on and start to key in commands. Using programs stored on the computer disk, you direct the computer to retrieve the stream of data on the tape and write it onto the computer screen.

If the instrument that you used was a spectrograph, you will see a graph showing the amount of light at each wavelength that reached the detectors. Using still other computer programs, you can search the data to find which features in the spectrum of the object correspond to spectral features measured in the laboratory. Having selected several such features, you can ask the computer to find the velocity shift that best fits all of the features, and you can thus determine and store for future reference the amount of Doppler shift (the wavelength changes caused by motion toward or away from us) in the spectrum of the object you are studying.

If, on the other hand, you were using either the wide-field or the faint-object camera, the tapes will show you a true image of the sky, as if you had "ultraviolet eyes." You may be looking at a faint galaxy never before imaged in detail. You may notice a new feature in the image, perhaps a small companion galaxy that appears to be somehow interacting with the main galaxy. Using the workstation's capacity to zoom in on small features, you may look more closely at the filament of gas apparently connecting the larger and the smaller galaxy. By activating other parts of the computer program, you can measure the brightness of the filament, its length and width, and its exact position in the sky. These data will be stored for later use.

Whether you are working with a spectrum or with an image, you will press other computer keys to print the data on a laser printer. Perhaps you will bring in an instant camera to photograph the image on the screen of your computer console for later publication in a journal.

As the weeks pass, more images and printouts will accumulate on your desk. You will try to figure out what they mean. Studying previous scientific articles that discuss the galaxy you have observed, you may see that it is unusual in certain ways. Perhaps the galaxy is known to emit x-rays. You realize that filaments of gas such as the gas you have observed have been seen before in x-ray galaxies, but that the one you have found is the longest and brightest ever seen.

Other scientists on your team participate in analyzing the Space Telescope's observations. Working via electronic mail from his or her home institution, the computer expert shows you how to process the image with special techniques that bring out the faintest details, and as a result you see that the filament spirals right into the center of the larger of the two galaxies. The theorist on your team suggests that the center of the larger galaxy could contain a supermassive black hole, already indicated by the telltale emission of x-rays. This black hole could be "swallowing" matter and thereby producing the energy observed as the infalling matter heats up through high-velocity collisions. Perhaps the filament you have discovered consists of gas being "stolen" from the small galaxy by the supermassive black hole in the big one! This situation has never before been observed, but your theorist digs out a paper in which an astrophysicist predicted on theoretical grounds that it should occur.

The team continually interacts by computer-borne electronic mail, working toward the draft of a scientific paper that all members of the team can approve. You assemble the arguments for and against your interpretation of the data. You finally agree to write a research paper about the galaxy you observed, propounding your new interpretation. In the paper, you agree, there will be a paragraph setting forth plans for new observations. You realize that if your interpretation is correct, the gas connecting the two galaxies should be moving at high velocity, and that the best way to check whether this is so is to obtain an ultraviolet spectrum with the faint-object spectrograph on the Space Telescope and to measure the Doppler shifts that should reveal any motion of the gas.

Tired but happy, you now know that your observing run on the Space Telescope was successful. You found out something new about the universe. The capture of gas from a companion galaxy may prove to be an important new phenomenon, explaining the energy source of black holes in galaxies and thereby clarifying our understanding of activity in the nuclei of galaxies.

It has been a long effort, not quite like the old days, when you would drive up the mountain to the observatory and obtain the data yourself. But thanks to the Space Telescope, you have seen something that no one in the history of the world could have possibly seen before.

SCIENTIFIC STYLE

Using the Space Telescope can be a complex business. It is "big science"—a phrase coined by Alvin Weinberg to describe scientific research done by large groups of scientists using large and expensive facilities. A prime example of big science is a particle accelerator or "atom smasher" like the Tevatron at the Fermi National Laboratory in Batavia, near Chicago, Illinois (Figure 39). Here hundreds

Figure 39: An aerial view of the Fermi National Laboratory, in Batavia, Illinois, reveals the great circle of the main accelerator ring of the Tevatron. Three lines tangent to it lead the beam of accelerated particles to halls where experiments are performed. *Fermilab*

of physicists and engineers operate the country's largest particle accelerator, capable of accelerating protons to within one part in a million of the speed of light, where their energy is equivalent to a temperature of ten million billion degrees.

It was not always thus. Before World War II, physicists worked in groups of two or three in small university laboratories, building experimental apparatus that fit into a single room and, along with graduate students, sometimes working at night, on Sundays, and on holidays to measure physical properties of things like atoms, molecules, and atomic nuclei. Usually these laboratories occupied rooms in physics buildings on college campuses, and the professor could reach them with a short walk from the parking lot.

During the Second World War, the Manhattan Project and the MIT Radiation Laboratory changed all that. Rallying to the call of such well-known figures as Arthur Holly Compton, James Conant, Vannevar Bush, Ernest O. Lawrence, and J. Robert Oppenheimer, physicists threw themselves into the war effort. Virtually overnight they organized sprawling laboratories at the Massachusetts Institute of Technology in Cambridge, Massachusetts, at the University of Chicago's facility in Argonne, Illinois, and at Los Alamos, New Mexico, where the largest of these laboratories, in the charge of the University of California, took shape. These laboratories had a single purpose—to develop, as quickly as possible, two devices that were to play crucial roles in winning the war: radar and nuclear bombs. The story of these heroic efforts is well known, while the impact of the bomb on postwar international politics is still being played out. World War II also taught physicists how to do big science.

After the war, many physicists saw opportunities to learn about the basic structure of matter—the nuclei of atoms—by working together in groups, using large particle accelerators. The work in nuclear physics needed to make the atomic bomb had whetted physicists' appetites to understand the forces that hold the atomic nucleus together.

What, for example, holds together the eight protons in an
oxygen nucleus, even though the electrical repulsion among
them would be expected to thrust them apart? What attractive force holds the protons together? How does that force
work? How does it relate to the other forces of nature, such
as electromagnetic forces and gravitational forces?

By chance, the research in microwave electronics that
had been needed for the development of radar put new
experimental tools into the hands of the physicists, supple-
menting the new instruments that had been developed in
the bomb laboratories. As a result, it became possible to
design new and more powerful types of particle accelerators, in which electronic devices would be used to accelerate small numbers of protons to high energies. Directed
toward targets made of specific chemical elements, these
energetic protons would collide with the nuclei of those
elements at tremendously high speeds, smashing them into
their component protons and neutrons. By observing how
other particles—mesons, for example—were produced in
the collisions, particle physicists could attempt to figure
out what holds nuclei together.

Before the war, the expense of such particle accelerators
would have ruled them out of consideration, but after the
war, particle physicists, accustomed to obtaining large
sums from the federal government, proposed that the government continue to fund their research, not for war, but
for science per se. Vannevar Bush, in his *Science, the
Endless Frontier*, argued that long-term investments in
science would prove as useful for the country in peacetime
as they had been in war. Congress approved substantial
funds, and the era of big science had begun.

In 1980, the Tevatron at Fermilab (as the Fermi National
Laboratory is called) was completed at a cost of $500
million; its operating budget is now $150 million per year.
Fermilab employs 2,100 people. That's what we mean by
"big science." No longer do all physicists simply drive to
their university laboratories. More likely, the particle
physicists among them board a plane for Chicago or even
Geneva, Switzerland (where the world's largest particle

accelerator is located). A joke among such physicists is that they are more likely to meet their faculty colleagues at O'Hare Airport than in the parking lot at home.

The planned follow-on to the Tevatron is the Superconducting Supercollider, or SSC, which is to be built in Waxahachie, Texas, south of Dallas–Forth Worth, during the 1990s at an estimated cost of $4 billion, eight times the cost of the Tevatron (Figure 40). Some of the increase in the

Figure 40: This satellite image of Washington, D.C., over which is superposed the plan of the future Superconducting Supercollider particle accelerator, conveys a feeling for the size of the new machine, to be constructed near Waxahachie, Texas. *Universities Research Association*

cost of particle accelerators arises from inflation, but nevertheless the SSC is an example not of big science but of superscience. Without examining the pros and cons of this giant project, we may note that people within and without the scientific community are asking, "Is it worth the cost? Where do we stop in the pursuit of knowledge about particle physics?" These are important questions, touching on the basic values of our society, as well as economic and political questions.

In this context, it is important to realize that the nearly $2 billion cost of the Space Telescope actually exceeded the cost of building the Tevatron, although the Space Telescope was less expensive than the SSC will be. A case could be made that the Space Telescope is the most expensive scientific instrument ever built. On the other hand, some scientists think that much of the telescope's cost stems from the fact that the instrument was designed to be launched into space and serviced in orbit by human beings using the Space Transportation System (the Space Shuttle). Because the decision to launch the Space Telescope using a manned vehicle was driven by a national goal to exercise manned space capability, one could argue that the entire cost of the Space Telescope should not be charged to science alone.

Like their physicist colleagues, astronomers before World War II worked almost entirely at colleges and universities. Many worked at observatories such as Mount Wilson, Lick, Yerkes, Lowell, and McDonald, all of which had been created with private endowments, a trend that had been established in the late nineteenth century, when astronomy was seen as a practical science worthy of support and long before federal funds for science were available. But like physicists, many astronomers also participated in the war effort and saw the opportunity afterward to secure government support.

First off the mark were the radio astronomers, a group of physicists, astronomers, and engineers who sought to apply the microwave technology developed for wartime radar to the study of radio waves from space. The National

Science Foundation, established in 1950, made some of its first grants for radio astronomy facilities at universities, among them the California Institute of Technology, Harvard, and the University of Michigan. But optical astronomers were not far behind in seeking government support, and soon reflecting telescopes of moderate size (thirty-six to forty-eight inches) were springing up around the country, funded in large part by the National Science Foundation.

Astronomers realized, however, that studying faint objects would require larger telescopes. While the particle physicists needed higher energies in order to probe deeper into the atomic nucleus, astronomers needed more light-gathering power in order to peer deeper into space. The astronomers followed the lead of the particle physicists, who had established a consortium of universities (Associated Universities, Incorporated, or AUI) to build a particle accelerator at the Brookhaven National Laboratory on Long Island. They requested that AUI establish the National Radio Astronomy Observatory (NRAO) in Green Bank, West Virginia (whose headquarters is now in Charlottesville, Virginia), to build much-desired radio telescopes. Then the astronomers established a new consortium, called the Associated Universities for Research in Astronomy (AURA), to build and operate the Kitt Peak National Observatory [now part of the National Optical Astronomy Observatories (NOAO) in Tucson, Arizona]— with major additional observing facilities at Cerro Tololo, Chile, and Sunspot, New Mexico.

A series of great radio telescopes developed by the National Radio Astronomy Observatory culminated in 1978 with the completion of the Very Large Array, twenty-seven radio dishes mounted on a Y-shaped system of railroad tracks that spread for more than fifteen miles near Socorro, New Mexico (Figure 41). The most powerful telescope of its kind in the world, the VLA can make images of even faint radio sources with fine resolution, less than one second of arc, in less than a day.

Figure 41: The Very Large Array of radio telescopes near Socorro, New Mexico. *National Radio Astronomy Observatory*

The National Optical Astronomy Observatories constructed two four-meter (160-inch) optical telescopes on Kitt Peak and at Cerro Tololo. Although Caltech's five-meter (200-inch) Hale Telescope on Palomar Mountain, California, is larger, these two new telescopes are well instrumented and are employed very efficiently by the entire astronomical community. The world's largest telescope, the ten-meter (400-inch) Keck Telescope, is now under construction at the Mauna Kea Observatory in Hawaii, supported mostly by private funds. Still, NOAO is planning to build a National New Technology Telescope at the same observatory, using four eight-meter (320-inch) mirrors to obtain even greater light-gathering power than the Keck Telescope will have. Each of the major national radio and optical observatory facilities produce some 400 research papers every year, touching every aspect of modern astronomy from planets and their satellites to stars, galaxies, quasars, and the universe.

By operating the National Observatories with federal funds provided by the National Science Foundation, astronomers learned how to cooperate in designing, operating, managing, and using larger instruments than any single university group could afford. With the success of its space astronomy programs, NASA realized that it too operates facilities that can be used in the same way as the National Observatories by the entire U.S. astronomical community. Like them, the functioning space facilities, which now include the Space Telescope, are open to any investigator who wishes to propose an observing project. When the Space Telescope became a reality, astronomers sought through AURA to establish a special facility to direct its scientific program; the Space Telescope Science Institute in Baltimore, funded by NASA and ESA, is the result. Hence AURA now manages the scientific operation of facilities in space as well as on the ground (Kitt Peak and Cerro Tololo). In this way, U.S. astronomy has become big science, and the Space Telescope, with its complex institutional arrangements, its large cadre of scientists, and its highly bureaucratic process for proposal review, has become the largest scientific project in all of astronomy, rivaling the particle accelerators of the physicists.

Big science has its drawbacks. Older astronomers look back wistfully to the days when a single person could journey to the observatory armed with his or her observing list, train the telescope on a distant galaxy, mount the photographic plate for a six-hour exposure, and then develop the plate in the observatory darkroom before going to bed at dawn. Descending the mountain after his or her several-night "observing run," the astronomer would pore over the data back at the office and discuss them over tea in the department library. After a suitable interval, the astronomer would publish them in the *Astrophysical Journal*, a permanent record of his or her efforts and findings.

A life of scholarship, of studying nature, aided by a precision instrument located on a beautiful mountaintop—what more could a person of science ask? Those days are gone forever, say the old-timers, and with a few exceptions, they are no doubt correct. High-tech, bureaucratic folderol, piles of magnetic tape, and a host of other scientists with a part in your project are the price that must be paid to use expensive national facilities like the Space Telescope.

Is the price worth it? If the past is a guide to the future, yes. There is simply no other way to obtain the sharper images, the access to fainter objects, and the ultraviolet spectra that are certain to yield key new information about a whole host of astronomical phenomena. Nevertheless, the question should be asked.

Will the Space Telescope find something fundamentally new about the system of the physical world that will justify its great expense? Faced by this question, even the most enthusiastic scientists must in all honesty admit their ignorance. They don't know, but as people of incurable curiosity, they are inspired by the example of their older colleagues, who, turning their telescopes to the skies during the 1960s, found objects as startling and interesting as quasars, enigmatic explosions in the depths of the universe.

They hope to be as lucky.

7

SCIENCE FROM THE SPACE TELESCOPE

THE PREVIOUS CHAPTERS have provided an overview of what the Space Telescope can do and how it does it. From a scientific perspective, these chapters have been like time in the bullpen; it is time now for the Space Telescope to take the mound and see if it truly has the right stuff.

ANTICIPATING THE UNEXPECTED

To discuss the astronomical discoveries to be made by the Hubble Space Telescope may seem premature, in view of the fact that the Space Telescope is just beginning its projected twelve- to fifteen-year life. (On the other hand, it must be admitted that if astronomers had nothing to say about the discoveries that they expect to achieve with the Space Telescope, a wise Congress might have been justified in withholding the funds to design, build, and launch it.) Yet history shows that each time a new, better instrument has been created to study the cosmos, entirely new and unexpected results emerge. Just what those marvels may be remains a question for the future. What we do know is that the Space Telescope can greatly improve astronomers' research into scientific issues now under investigation and solve some of the mysteries that are already apparent.

The general astronomical areas most in need of the Space Telescope's abilities may be separated into four categories. First, with the Space Telescope, we can study the planets

159

in our own solar system continuously, with a sharpness of view comparable to that obtainable from spacecraft that spend only a few days in the planets' neighborhoods. Second, we can attempt to find planets—if they exist!—around stars close to the sun. If successful, this search would increase our confidence that the sun's planets are not a cosmic anomaly, and would thus increase our estimates of the number of likely sites where extraterrestrial life could exist. Third, we can study the distribution and composition of interstellar gas and dust in the Milky Way far better from space than we can from Earth-based telescopes. Fourth and most important—at least in the authors' opinion—we can study cosmology, the science of the universe as a whole, and can perhaps resolve certain ultimate mysteries that have been perplexing astronomers for the past fifty or sixty years: Will the universe expand forever? Is the universe finite or infinite in extent? Do the stars and galaxies that we see represent most of the universe, or does most of the matter in the universe reside in an invisible form, currently unknown to us?

THE CONTINUING IMPORTANCE OF EARTH-BASED OBSERVATORIES

The advent of data from the Space Telescope will not mean that the great ground-based observatories on Earth— Palomar Mountain in California; Kitt Peak and Mount Hopkins in Arizona; Cerro Tololo, La Silla, and Las Campanas in Chile; Mauna Kea in Hawaii; and a host of others— will go out of business. On the contrary, the Space Telescope will make these observatories even more important than they are today.

Why? The answer lies in the limited amount that the Space Telescope can accomplish, even with its tremendous abilities. To be sure, the Space Telescope can obtain images of galaxies with fantastic detail and can study features in the ultraviolet spectra of stars that can be observed in no other way. But with 5,000 scientists waiting to use it for almost as many projects, most of them will receive no more than a few hours a year with the Space Telescope. This enormous demand for time implies that the Space Telescope should be used only for projects that can be accomplished in no other way. But precisely because these projects are unique and limited in number, the results from the projects will stimulate a host of new hypotheses to be investigated with ground-based telescopes.

One example of this ripple effect will arise in spectroscopy, the detailed study of the radiation from a planet, star or galaxy at different wavelengths. Suppose that the Space Telescope observes ultraviolet radiation from a faint star and finds that the star's spectrum is highly unusual, perhaps indicating that the star contains technetium—a radioactive element. You can bet that ground-based telescopes will soon thereafter be pointing at that star, gathering information in the visible part of the spectrum to test and extend the findings of the Space Telescope.

Or consider the case in which the Space Telescope observes a quasar—an energetic object in the depths of the universe believed to harbor a black hole. While the telescope is gathering unique information about the ultraviolet spectrum of the quasar, its wide-field camera records the image of the surrounding sky. Reviewing these data in the archives, an astronomer notes that dozens of very faint galaxies appear in the vicinity of the quasar. The question then naturally arises as to whether the galaxies are actually associated with the quasar in *space*, or happen to lie coincidentally along the same line of sight at some intermediate distance between ourselves and the quasar. One way to determine the answer is to measure the amount of the Doppler shift of the galaxies and then compare this amount with that for the quasar. A project to make this measurement with the Space Telescope would not be approved by the committee, because a large ground-based telescope can make these measurements in visible light. So proposals to measure the Doppler shifts with ground-based telescopes will emerge.

A final example involves observations of the sun's planets. Suppose that astronomers predict that one of Saturn's satellites will undergo an occultation at a particular moment—that is, it will suddenly disappear behind its parent planet as it moves along its orbit. Ground-based telescopes can observe the occultation in visible light and find how light is absorbed by Saturn's atmosphere on its way toward us. But only the Space Telescope, free of Earth's atmospheric blurring, can pinpoint the exact position where the occultation occurs. For this reason a planetary scientist working with ground-based telescopes might request a brief observation by the Space Telescope. But the bulk of the observation will still be performed with ground-based telescopes.

It is a happy fact that the large ground-based observatories and the Space Telescope complement one another; that is, each performs essential work that the other cannot. The Space Telescope excels at obtaining the sharpest possible images of the cosmos and observing the universe in the ultraviolet, a spectral region completely inaccessible to ground-based observatories. The large telescopes on high mountains excel at another task—collecting the largest amount of visible light possible, which they perform especially well because of the large diameters of their mirrors. They also allow astronomers to collect data on a host of objects that the Space Telescope will be unable to observe during its limited lifetime. In short, there is only one Space Telescope, with its marvelous functions, but there are also, thankfully, a host of giant ground-based telescopes.

CHOOSING PROJECTS FOR THE SPACE TELESCOPE

It will be up to the astronomical community to maximize the usefulness of the Space Telescope by selecting only the most appropriate projects, relying on ground-based telescopes to carry out observations that do not require the Space Telescope's unique capabilities or that cannot be scheduled because other, better projects have already preempted the available time. How can the usefulness of the Space Telescope be maximized?

In an attempt to answer this question, the Space Telescope Time Allocation Committee has adopted a policy of assigning one-third of the available observing time to "small" projects, one-third to "medium-sized" projects, and one-third to "large" projects. The reason for doing this is that experience with large ground-based telescopes has shown that when a time allocation committee tries to satisfy the most astronomers, there is a tendency to divide the time available into smaller and smaller slices. Recognizing that certain projects are so large that they will never get finished with such an approach, the Space Telescope Time Allocation Committee attempts to remedy this problem by assuring that medium-sized and large projects will each receive a substantial fraction of the time.

Among the large projects, the Space Telescope Science Institute has established a special category of "key projects." A key project must satisfy three requirements: The committee must judge that it would be a disaster if the Space Telescope did not perform such a project. The project must require the efforts of a team of astronomers with a variety of areas of expertise, not of individuals working alone. And the project must include coordination of dissenting views about how the project should be undertaken, in order to reflect concerns of the entire astronomical community, not only those of the key scientists involved in the project. A perfect example of such a key project lies in the cosmology research described in the next chapter.

PLANETARY ASTRONOMY WITH THE SPACE TELESCOPE

A prime area of observation for the Space Telescope will involve detailed study of the planets in our solar system. (Extrasolar planets, which might someday be detected by the Space Telescope, can hardly be planned subjects of

Figure 42: *(Top)* An image of Jupiter and its satellite Io obtained by *Voyager 2* on June 25, 1979. *(Bottom)* Saturn and its rings as seen by *Voyager 1* on October 30, 1980. *NASA, Jet Propulsion Laboratory*

observation, and in any event would fall in the category of stellar observations.) Here the Space Telescope's amazing ability to obtain crisp images operates to the fullest advantage. Even though spacecraft such as the two Voyagers have traveled close by the planets, reaching distances from Jupiter and Saturn that are only one ten-thousandth of the Earth's distance from these giant planets, the Space Telescope will be able to "see" Jupiter and Saturn almost as well as the Voyager spacecraft did (Figure 42).

The reason for this lies in the Space Telescope's large primary mirror and its advanced optical system; in contrast, because of weight limitations, the Voyager and other spacecraft could carry only cameras with small lenses. Because they flew close by the planets, these spacecraft could obtain sharp images even without a huge lens or mirror. The Space Telescope's fine optics allow the near-duplication of the Voyager observations without leaving the vicinity of the Earth!

One might ask why we should "duplicate" previously made observations. The answer lies in something that the Space Telescope has but the Voyagers did not: time.

As astronomers have increasingly come to recognize, the universe is not a static place, of which a single glimpse will suffice to provide a true picture. Instead, change and evolution occur constantly on all scales. Some processes, such as the collapse of a stellar core to produce a supernova explosion, require one second or less. Other processes, such as the weather on Earth, occur on time scales measured in hours or days. So do changes in the energy output of Cepheid variable stars, which pulsate because of an internal instability. Still other processes, such as the long-term climatic changes on this planet, or the changes in energy output from a quasar, are measured in years or centuries, even in millennia. And other events require tens or hundreds of thousands of years to occur (the late phases of a star's evolution, for example), if not many millions of years (such as changes in a star during the prime of its life). If you want to understand the universe fully, you must match your observational program to the rhythms of nature, as

Figure 43: Jupiter's Great Red Spot and the counter-rotating cyclone pattern below it, photographed by *Voyager 2* on July 3, 1979. *NASA, Jet Propulsion Laboratory*

far as that is possible. You cannot, for example, hope to understand the Earth's weather patterns by taking a single snapshot of cloud patterns.

So it is with the sun's largest planets, Jupiter and Saturn. Astronomers have long been fascinated with the changing "bands" of color on the two planets, especially on Jupiter, the larger (eleven times the Earth's diameter as opposed to nine and one half times for Saturn) and closer (five times the Earth-sun distance as opposed to nine times for Saturn). The most prominent of Jupiter's banded features, which have been known since Christian Huygens studied Jupiter during the late seventeenth century, is the famous Great Red Spot. In addition, swirling around the Great Red Spot are changing colored bands—red, orange, yellow, and brown—that run parallel to the planet's equator (Figure 43).

Astronomers are now convinced that these colored bands are the top of Jupiter's visible atmosphere, in which both rising and falling currents are moving, superimposed on the general west-to-east pattern of atmospheric circulation that is reminiscent of the Earth's atmosphere, but far stronger. Jupiter rotates in far less time than the Earth does (ten hours rather than twenty-four), and since the giant planet has eleven times the Earth's size, this rapid rotation means that Jupiter's equatorial regions are moving about twenty-five times faster than those on Earth—25,000 miles per hour at the equator instead of 1,000. Such rapid motion can drive a circulation pattern that puts Earth's to shame.

The Great Red Spot, for example, is a sort of supercyclone, a pattern of swirling gas that has persisted for at least three centuries, ever since Huygens first noted it, and perhaps much longer than that. However, the position of the Great Red Spot is not fixed. Instead, the cyclone wanders somewhat through the atmosphere, additional proof that what is observed is the motion of atmospheric gases.

Close to the Great Red Spot, a "countercyclone," much paler than the Red Spot itself but perfectly visible on the Voyager photographs, moves in tandem with the Red Spot. The gases in this countercyclone are rotating in the direction opposite to the motions of the gas within the Great Red Spot itself.

Scientists would love to obtain time-lapse observations, hour by hour, day by day, year by year, in order to study how and why Jupiter's atmospheric patterns change. One of the greatest legacies from the two Voyager flybys of Jupiter is a sort of movie, a montage of photographs extending over a day and a half as the spacecraft neared and passed the giant planet. The images were properly rescaled to reflect the changing apparent size of the planet that arose from the varying distance of the spacecraft from Jupiter. These images were then photographed one by one to produce a movie, a tantalizingly brief look at the weather on Jupiter. For scientists, this is something like having a day and a half of Earth's weather to study in an attempt to determine what our weather patterns are and what causes them.

We should not underestimate our ignorance of Jupiter's atmosphere, far though we may have come since the days of Huygens. Take, for example, those marvelous colors, which make Jupiter reflect sunlight in various hues. What causes these colors? We know that Jupiter's atmosphere, like most of the universe, consists mainly of hydrogen and helium. The hydrogen atoms have paired, for the most part, to form hydrogen molecules (H_2), whereas the helium atoms have paired with nothing, as is their wont. But neither hydrogen molecules nor helium atoms produce shades of red, orange, yellow, or brown when they reflect light; instead, these gases are essentially colorless. Something other than hydrogen or helium must be present in relatively small quantities to produce the vivid coloration. But what?

Astronomers don't know; that is, some claim to know, but these claims disagree. In fact, there is no agreement as to whether the compounds that produce Jupiter's coloration are organic in nature (based on carbon atoms) or inorganic (with carbon absent), a fundamental distinction made for molecular compounds of atoms on Earth. The answer may lie in the ultraviolet spectrum, which the Space Telescope is well suited to study. Atoms and molecules that have few "fingerprints" in the visible spectrum often leave evidence in the ultraviolet. If so, the Space Telescope will find it.

So it should come as no surprise that astronomers don't know the details of Jupiter's weather patterns and what causes them to occur, save for the fundamental thought that Jupiter's rapid rotation must lie at the heart of things. Here there is no substitute for observational data, and such data are just what the Space Telescope can provide. A planetary astronomer might well desire that the Space Telescope spend half its observing time obtaining pictures of Jupiter. Fortunately, Jupiter is so bright that obtaining an image will take a very short time, so even a repetitive series of time-lapse images for use in making a movie will require that the telescope spend only a few percent or so of its time observing Jupiter. But even this small amount of time will provide a series of observations of Jupiter hundreds of times more extensive in time than those sent back by the Voyager spacecraft. Ground-based telescopes cannot help. Atmospheric blurring eliminates any hope of recording the intricate patterns in the clouds. Our largest ground-based telescopes can, like Huygens's, observe the appearance of a banded pattern, but they are unable to provide much additional information.

When astronomers examine the atmospheric patterns—the "weather"—on Jupiter, they are after more than observational data concerning the largest of the sun's planets. The global view of Jupiter is, in a way, superior to anything we can obtain of our own planet, whose weather patterns are not clearly marked by vividly colored bands of material. A photograph of Earth therefore misses some of our actual atmospheric motions. From time-lapse observations of Jupiter, scientists hope to reconstruct the entire global weather picture on that planet—a picture that, by analogy, may help to explain the Earth's overall weather cycles.

What goes for Jupiter holds true, to a lesser degree, for Saturn. Saturn too has banded regions parallel to its equator, though the colors of these bands are far less intense than those on Jupiter. Like Jupiter, Saturn rotates in about ten hours, and since Saturn is far larger than Earth, Saturn likewise rotates far more rapidly at its equator than Earth does. The Space Telescope will also observe Saturn (Figure 44), though Saturn will be a more difficult subject than Jupiter because of Saturn's greater distance from the Earth and the relatively greater difficulty of observing its weather patterns.

Uranus and Neptune, the two even more remote of the sun's giant planets, will receive still less attention, though they too will be observed with far greater clarity than any ground-based observatory can provide. Uranus has nine-

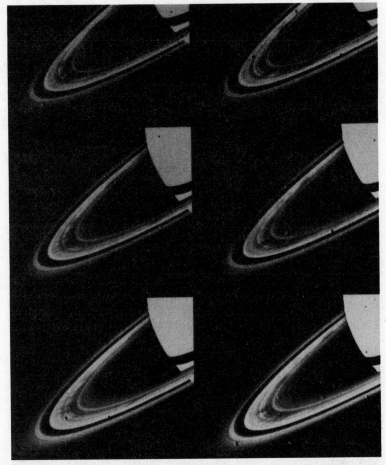

Figure 44: *Voyager 1* images of Saturn's rings taken on October 25, 1980, showing changing, spokelike patterns in the distribution of the tiny particles that form the rings. *NASA, Jet Propulsion Laboratory*

teen times the Earth-sun distance from our planet, and Neptune has thirty times this distance. Hence our view of these two large planets (each has about four times the Earth's diameter) will be less sharp than our observations of the two inner, larger planets.

In addition to studying the giant planets themselves, the Space Telescope will be able to observe the surface features of the larger satellites of these planets, in particular

Jupiter's four large satellites, each as large as or larger than our own moon. Once again, although the Voyager spacecraft took excellent photographs of these objects, astronomers would dearly love to observe the changes that occur with time, especially on Io, the innermost of Jupiter's four large satellites (Figure 45).

Io (pronounced eye-oh or ee-oh at whim) is the only object in the solar system other than Earth known to have active volcanoes. These volcanoes continually deposit sulfur-rich compounds on the satellite's surface, changing its features and spewing noxious fumes to heights of several

Figure 45: Io, Jupiter's innermost large satellite, observed by *Voyager 1* on March 4, 1979. Many of its surface features, including the doughnut-shaped feature in the center, are caused by volcanoes. *NASA, Jet Propulsion Laboratory*

hundred miles. Time-lapse studies of Io's volcanic activity promise to reveal a great deal about what causes these volcanoes and how they affect the surface of Io over the years.

The Space Telescope will also observe the largest of Saturn's satellites, Titan and Rhea, and may possibly discover the existence of small unknown satellites around Saturn, Uranus, and Neptune that cannot be seen with ground-based telescopes, should they exist. (As for Jupiter, astronomers generally believe that the two Voyager spacecraft, which discovered half a dozen new satellites, probably completed the task as well as the Space Telescope can.) Still more exciting could be the hunt for a possible tenth planet of the sun, far beyond the orbit of Pluto, which moves in a frigid path at forty times the Earth-sun distance. The search for a possible trans-Plutonian planet, whose existence has been suggested by the motions of Neptune and Pluto, which may reflect such a planet's gravitational pull, has extended over two decades, but so far without success.

In addition to its observations of the giant planets and their satellites, the Space Telescope will also obtain an ongoing record of the rings of the giant planets. Although Saturn's ring system is by far the widest and most famous in the solar system, automated spacecraft passing by Jupiter and Uranus have verified that they too have rings, though narrower and hence harder to see from Earth. The Voyager encounter with Neptune in August 1989 revealed that Neptune also has rings, so all four giant planets turn out to be surrounded by ring systems.

All of these ring systems consist of countless small particles, ranging in size from boulders down to microscopic dust grains, that circle the planet in smaller orbits than those of its larger moons. Although the ring particles are each quite small and therefore exert only tiny amounts of gravitational force on one another, they nevertheless interact collectively in intriguing ways. For example, in Saturn's rings are many thousands of individual orbits in which the ring particles tend to congregate. In some cases, it is obvious that a particular orbit has become heavily populated with particles because a particularly large particle, that is, a visible satellite of Saturn, is shepherding particles into that orbit with its gravitational force.

The Voyager spacecraft that sailed past Saturn in 1980 and 1981 discovered another aspect of the collective interactions among the ring particles. Seen face on, rather than on edge, the rings are not uniform, but instead show complex, spokelike patterns. This means that some areas within the rings contain more particles than average, and some fewer than average. These patterns arise from the complex, mutual interactions of the billions of particles that form the rings. Using the Space Telescope, astronomers will obtain a series of images of the ring patterns, primarily of Saturn's rings but also of Jupiter's and Uranus's, that may reveal the particle dynamics that produce these patterns. Even our best computers are inadequate to model the interactions among these billions of particles, but the Space Telescope's observations, extending over a period of time, may suggest simplified models that explain the changes in the ring patterns.

The sun's family includes not only the sun, its nine planets, and their satellites, but also a host of smaller objects called asteroids, as well as comets—lumps of ice and dust left over from the formation of the solar system. Asteroids are rocky objects, ranging in size from 400 miles across for the largest, Ceres, down to a mile or less for the smallest yet discovered. Doubtless a host of asteroids smaller than this exist, still below our threshold of detectability. Most of the asteroids orbit the sun between the orbits of Mars and Jupiter. They apparently represent the material for a planet that could never form because the perturbing effects of Jupiter's gravitation kept it from coalescing.

Today the asteroids remain a great mystery. We have yet to observe any asteroid's surface directly, unless we count the observation by the Viking orbiter of the surfaces of

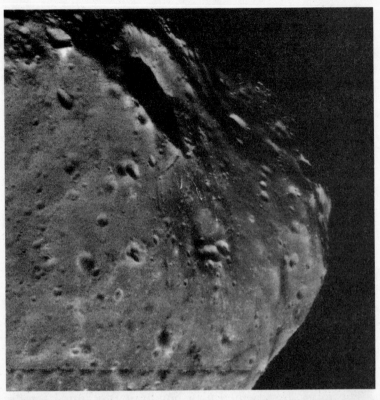

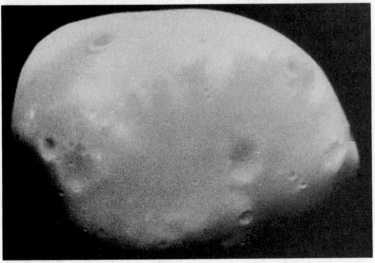

Figure 46 (opposite): Mars's two small satellites, Phobos *(top)* and Deimos *(bottom),* are probably captured asteroids. These images were obtained by the *Viking* orbiter spacecraft. *NASA, Jet Propulsion Laboratory*

Mars's two tiny moons, Phobos and Deimos, each about a dozen miles across, which are probably asteroids captured by Mars's gravity (Figure 46). The Space Telescope will, for the first time, be able to get a good look at an asteroid's surface, since the telescope will be able to see details about twenty miles across on an object between Mars and Jupiter.

STELLAR ASTRONOMY WITH THE SPACE TELESCOPE

The stars that shine in the night skies have an interesting property that has received little attention on Earth: some of them emit more energy each second in the form of ultraviolet radiation than they do in the form of visible light. Our own sun, a typical star, does not conform to this rule. The sun's ultraviolet energy—blocked from Earth by our life-giving, protective atmosphere—amounts to only about 10 percent as much energy per second as the amount that the sun emits as visible light, which does penetrate our atmosphere and which our eyes detect as sunlight.

The fact that many stars emit copious streams of ultraviolet suggests that if you seek to understand stars in detail, you ought to examine their ultraviolet as well as their visible-light radiation. Stars hotter than the sun emit a larger fraction of their energy in the form of ultraviolet; the hottest stars produce nearly all of their radiation as ultraviolet. Furthermore, many of the most abundant chemical elements in the universe—for example, helium and carbon—produce many absorption lines (wavelengths at which the radiation has been removed) in the ultraviolet region of the electromagnetic spectrum, but almost none at all in the visible-light portion of the spectrum. Therefore, in the case of several of the most important types of stars, if

you hope to determine what elements the stars contain and
in what amounts, you barely advance your knowledge if
you confine yourself to the visible-light portion of the
spectrum that can be studied with ground-based observatories. What you need is to place a spectrograph above the
atmosphere, capable of recording ultraviolet absorption
lines.

Astronomers have already done this, first with the *Copernicus* satellite and then with the *International Ultraviolet Explorer (IUE)* satellite. The *IUE* still moves in its
synchronous orbit, a fine astronomical observatory, 22,000
miles above the atmosphere. But the former success of the
Copernicus satellite and the continuing success of the *IUE*
have simply whetted astronomers' appetites for more and
better ultraviolet spectral observations. Only with such
improved observations can they hope to discover the exact
elemental composition of stars' outer layers.

This may seem a sort of nit-picking, stamp-collecting
desire, the urge to know everything to the second or third
decimal place, but far more than this rides on accurate
observations of the exact abundances of the elements in
stars. Accurate determinations of abundances are the key
to understanding the synthesis of chemical elements in
stars. For years, physicists have predicted that heavy
elements such as carbon, nitrogen, oxygen, magnesium,
iron, gold, and uranium are created in the intense heat of
supernova explosions, and Supernova 1987A has finally
confirmed this by direct observation. But for years astronomers have found indirect evidence of the buildup of heavy
elements in the Milky Way by studying the abundances of
elements in stars; the more recently the star formed from
the interstellar medium, the more heavy elements it has.
Just which elements and just how much of each are the
keys to this study, and many elements reveal themselves
only through the absorption lines they produce in the
ultraviolet region of the spectrum.

But the Space Telescope offers astronomers more than
the chance to discriminate among various schemes by

which supernova-produced chemical elements affect the evolution of the Milky Way. Ultraviolet observations are particularly sensitive in detecting high-temperature gas. They can readily reveal whether any star, even a relatively cool one, possesses a "chromosphere," a layer just outside a star's surface that exists only because shock waves originating in the bubbling motions of the atmosphere below heat the gases nearby, or because electrical currents associated with stellar magnetic fields perform such heating. As we explained in Chapter 2, observations of the sun in x-rays revealed that it has a gaseous corona at a temperature of four million degrees. Ultraviolet observations of the sun reveal its chromosphere, which, like the corona, undergoes changes in the amount of its activity throughout the eleven-year magnetic cycle.

IUE ultraviolet observations have revealed the chromospheres of other stars, and Space Telescope observations will allow the chromospheres of fainter stars to be studied in an effort to find the underlying cause of magnetic activity in stars. This challenging scientific question has a practical side as well, because ultraviolet radiation from the chromosphere plays an important role in ionizing the outer parts of the Earth's atmosphere and thus affects radio communications on Earth.

One of the current frontiers of astronomy involves the search for planets that may exist around other stars. Until we find such planets, we cannot determine whether our solar system represents a unique, or nearly unique, anomaly in the sense that nearly all stars may lack planetary systems. Because extrasolar planets would shine only faintly by their reflected starlight, and because the star itself shines much more brightly than the planets do, we must attempt indirect detection as well as direct observation of planets around other stars.

The straightforward way to detect extrasolar planets is to see them directly. To aid in this attempt, the faint-object camera aboard the Space Telescope has a "coronographic finger," a thin rod that projects into the field of view to block the light from a star so that its much fainter planets may be seen. This method depends, of course, on precise guiding of the Space Telescope so that the coronographic finger remains precisely in front of the star yet does not block the light from the planet nearby. If this precision requirement is met, the Space Telescope should be able to see Jupiter-sized planets around the hundred closest stars—if the planets are there to be seen.

In addition, we can search for extrasolar planets by indirect means. One such method relies on studying the effect that the planets' gravitational forces must produce on their parent stars. No star remains entirely still in space; each has its own motion, typically a nearly circular orbit around the center of its galaxy. Our sun, for example, moves around the center of the Milky Way in a nearly circular orbit, taking 240 million years (sometimes respectfully referred to as one cosmic year) to perform a complete orbit (Figure 47). Other stars likewise move in their own orbits, not perfect circles but nearly so. We observe these

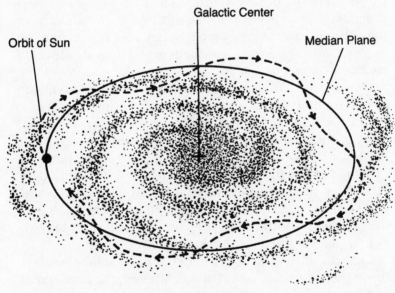

Figure 47: The sun's orbit around the center of the Milky Way. This drawing exaggerates the bobbing, up-and-down motion.

stars from a planet that moves in a tremendously smaller, more or less circular (actually elliptical) orbit around its star, as the star itself orbits the galactic center in an orbit about two billion times larger than the Earth's orbit around the sun!

If we take account of the parallax effect—the small changes in stars' apparent positions on the sky that arise from the Earth's motion in its yearly orbit around the sun—we can determine the motions of stars with respect to the center of mass of our solar system. Nearby stars are moving on orbits that are nearly but not quite parallel to the sun's orbit around the center of the galaxy. As a result, these stars have some apparent motion with respect to the sun. This motion reveals itself when we photograph a region of the sky twice, at intervals of many years. On the one hand, the distant stars all show the same relative positions, because their great distances make them appear motionless. On the other hand, the closest stars—especially those whose orbits around the Milky Way's center differ markedly from being circular—show positions that differ noticeably for the two times of observation. Over an interval of years or centuries, the paths of these stars appear to be straight lines, because we are looking at only a tiny fraction of a star's orbit around the galactic center, an orbit that takes hundreds of millions of years to complete.

A straight path will occur if the star moves by itself, not subject to the gravitational pull of any nearby objects. But suppose that the star has a planet in orbit around it. Although the planet does most of the moving—because it has far less mass than the parent star and therefore responds far more readily to a given amount of force acting upon it—it exerts a gravitational attraction on its star, in the same amount (but in the opposite direction) as the star exerts on it. This gravitational force makes the star move around its own little orbit.

In scientific terms, the center of mass of the star-planet system is not exactly at the center of the star. Instead, because the center of mass is the point that always lies between the star and its planet, and whose distance from the centers of these objects is inversely proportional to the masses of the star and planet, the center of mass always lies close to, but not precisely at, the center of the star. It is the center of mass that orbits the galaxy in a nearly circular path, part of which looks like a straight line.

The hypothetical planet performs a relatively huge orbit around the center of mass, while the star, always lined up on the other side of the center of mass, has a relatively tiny orbit. Tiny though it may be, we can hope nevertheless to detect the star's motion through careful observation. Because the Space Telescope produces very sharp images of stars, it will allow extremely accurate measurements of the motion of a star as it moves across the sky, including any deviations from straight-line motion. These tiny deviations may reveal planets around other stars through the effects of the planets' modest gravitational pull on their parent stars.

If a star has more than one planet, the situation grows more complex but not entirely hopeless. One of the planets is likely to be dominant in terms of gravitational force, as Jupiter is among the sun's family of nine planets. Jupiter, with 318 times the Earth's mass, exerts far more gravitational force on the sun than any other planet does; Saturn comes next, but since its mass equals "only" 95 times the Earth's mass, and since Saturn is nearly twice as far from the sun as Jupiter, Saturn's gravitational force on the sun equals barely one-sixteenth of Jupiter's. All the other planets, our own Earth included, have a much smaller effect on the sun's motion than either Jupiter or Saturn does.

An astronomer situated on a planet a few dozen light-years away—say, around the star Arcturus—could measure the sun's motion and deduce that the sun has one planet with approximately Jupiter's mass that orbits the sun every twelve years or so, for the observer would see periodic deviations in the sun's motion that repeated every twelve years as Jupiter completed an orbit. The observer could probably also determine that the sun has a second, less massive, and more distant planet that orbits the sun every twenty-nine years. If the observer had capabilities roughly similar to our own, the other planets would remain undetected. This method of finding planets is indirect but valid.

Although most stars—our own sun included—fade away quietly when they exhaust their supplies of nuclear fuel, some stars explode violently as supernovas. One of the most exciting astronomical events of recent years was the discovery of such an exploding star in the Large Magellanic Cloud, the closest galaxy to our own Milky Way (Figure 48). Supernova 1987A, detected on February 23, 1987, had in fact exploded about 160,000 years earlier; our knowledge of the cosmos is always a bit out of date because it takes a while for light to reach us.

Figure 48: Supernova 1987A (at the tip of the arrow) appeared in the Large Magellanic Cloud on February 23, 1987, indicating that a star had exploded. *National Optical Astronomy Observatories*

Supernovas are important not only for themselves, as interesting phenomena in a violent universe, but also for their effects on us. Every molecule in our bodies contains atomic nuclei that were made in exploded stars and later blasted into space to mingle with the other atoms and molecules there, perhaps later to form a new generation of stars and planets around them. For this reason as well as from natural curiosity, astronomers seek to find out more about the explosions that spread "evolved" matter among the stars.

One key way to do this is to follow a young supernova as it pushes matter farther into space. By "young," astronomers always mean young as we see it; thus, for example, Supernova 1987A is still quite a young supernova, 160,000 years old though it may be. Supernova 1987A is so young that, in astronomical terms, it has barely begun the expansion of its outer layers that will eventually add "mulch" to interstellar matter. A good supernova to study, if you seek to understand this mulching, is one that is a few hundred or a few thousand years old.

Astronomers know of several dozen such "supernova remnants," masses of gas and dust left behind by bright "new" stars. Ancient astronomers in China and the Middle East, as well as more modern astronomers in Europe, recorded the appearance and disappearance of these new superstars; today, when we look at the positions on the sky where the star once flared brightly, we find filamentary webs, thought to be the stars' outer parts, blown outward in the fury of the explosion. The Space Telescope offers us the chance to follow the evolution of supernova remnants, to see how the star-blown matter gradually merges with the rest of the interstellar medium.

THE INTERSTELLAR MEDIUM

Spread among the stars in the Milky Way galaxy, and occupying a volume fantastically larger than those tiny volumes filled by the glowing stellar globes, is the interstellar medium—gas, dust, and larger particles. This medium represents the cosmic nursery from which all stars were born, as well as the repository of material that has cycled through the stars that exploded and is now ready to form part of a new generation of stars.

Study of the interstellar medium has proceeded only slowly, hampered by a key characteristic of that medium: it does not usually emit much light. To be sure, we could "see" it if we had radio eyes. Certain types of atoms (most notably hydrogen) and molecules (most notably carbon monoxide) naturally emit radio waves at certain well-defined radio wavelengths. By studying the amounts of this radiation that reach the Earth from various directions, astronomers have mapped out the interstellar medium, at least for the denser parts of interstellar space—the hydrogen-containing (just about all) and carbon-monoxide-containing regions.

This is all a tremendous advance over the era before radio astronomy, when we were barely certain that interstellar matter exists at all (Figure 49). But if we seek to improve our knowledge, we already know that a powerful

Figure 49: A visible-light photograph of the Milky Way in Cygnus. The bright patches are due to emission by myriads of stars, while the dark lanes arise from the absorption of light by interstellar dust. *Mount Wilson and Las Campanas Observatories*

way to study the gas between the stars lies in making detailed ultraviolet observations of the spectra of stars. These spectra show absorption features—the removal of photons at certain frequencies or wavelengths—that arise from two distinct, though similar, causes. The first of these is the removal of photons in the outer layers of the stars themselves, as the starlight breaks through the denser regions below and sets forth on journeys of thousands of light-years. The second is the similar removal of photons by particular types of atoms, molecules, and ions floating freely in interstellar space, far from the star being observed. If such matter happens to lie along the line of sight between ourselves and a star, we can see the effect of the removal of photons by observing a drop in brightness, highly confined in wavelength, that appears when a particular type of atom or ion removes photons of a particular wavelength.

One might think at first that we would encounter great difficulty in determining whether such an absorption feature has arisen in a star's outer layer or in the interstellar medium, far from the star itself. In practice, however, this discrimination occurs rather easily. Once again the Doppler effect plays a key role. The atoms and ions that block or absorb the photons are not entirely motionless; they have random motions because their temperature lies above absolute zero. The higher the temperature, the greater the random velocities of the absorbing atoms and ions. Greater velocities produce a greater Doppler effect, which reveals itself as a "smearing out" of the absorption line, so that the intensity of starlight does not drop straight down at one and only one wavelength, but instead appears in diminished quantity over a band of neighboring wavelengths, with the greatest amount of absorption at the exact wavelength at which all the absorption would be concentrated if the temperature were indeed absolute zero. Hence, when astronomers study an absorption line in detail, they can determine the temperature of the atoms or ions that produced the blockage of starlight. Hotter absorbers produce a broader absorption line.

The outer layers of the stars hot enough to emit ultraviolet—the region of the spectrum best suited for detecting interstellar atoms, ions, and molecules—typically have temperatures measured in tens of thousands of degrees on the absolute scale, whereas the temperature of the cooler gases in interstellar space varies between 10 and 300 degrees above absolute zero, with temperatures of 20 to 100 degrees the most common. At these low temperatures (equal to −440 to −300 degrees Fahrenheit), an absorption line produced in interstellar space appears far sharper, far more concentrated toward a single wavelength, than does an absorption line that arises in a star's outer layers. As a result, astronomers can tell almost at a glance whether a given line is caused by gas located in the outer layers of the star under observation or in the much colder interstellar medium. They are ready to use this ability to study interstellar absorption lines in order to determine which types of atoms, ions, and molecules, how much of each type, and what temperature for each type characterize various regions of interstellar space in the Milky Way.

Once again, the most useful information is derived not from studying the visible light that passes through the interstellar medium, suffering some absorption at particular wavelengths as it does so, but from observing this behavior for photons of ultraviolet. This is so simply because atoms, ions, and molecules are so constructed that most of them can more readily produce absorption features in the ultraviolet than in the visible-light regions of the spectrum.

Thus, if we hope to understand the basic facts about the interstellar medium, we need an ultraviolet spectrograph, carefully observing the spectra of stars to see which absorption features appear, and which appear because of absorption by interstellar atoms and ions. This information may tell us what the medium is made of, how much of each element it contains, and what the temperatures in its various parts may be.

We can also hope to determine the velocities of clumps or clouds of interstellar matter along our line of sight, once again by using the Doppler effect. Such studies predate the Space Telescope, but they remain in relative infancy. The *Copernicus* satellite provided valuable information on the composition of interstellar matter, observing for the first time absorption features produced by a dozen different elements, all of which produce no absorption lines in the visible region of the spectrum. But the Space Telescope has a spectrograph whose accuracy is greater than that on the *Copernicus* satellite, and a telescope that is far more powerful than *Copernicus*'s.

We may therefore confidently expect that the Space Telescope will tell us far more about the details of the composition and temperature of interstellar matter than we know now. Such details will shed new light on perhaps the most exciting aspect of interstellar studies: the question of how and why, deep within relatively dense clouds of interstellar matter, stars begin to form.

Figure 50: M51, the Whirlpool Galaxy, is at a distance of about 10 million light-years. *Lick Observatory*

EXTRAGALACTIC ASTRONOMY WITH THE SPACE TELESCOPE

In addition to the cosmological studies described in the next chapter, the Space Telescope will provide astronomers with their best look ever at galaxies—a view so crisp and clear that it is hard to imagine what it will be like. Instead of merely seeing the broad outlines of the spiral arms of a galaxy such as the prototypical M51, the Whirlpool Galaxy, along with some of the details of those spiral arms, astronomers will be able to follow in intricate detail the whorls, swirls, and star clusters that define these arms (Figure 50).

Once again, this is not simply a question of obtaining a better catalog of the cosmos, however useful that may prove to be. Spiral galaxies currently demand close attention from astronomers on several grounds that the Space Telescope may help to resolve: How do such galaxies form and maintain a pattern of spiral arms? How did they ever form as galaxies in the first place? And why do some galaxies develop these characteristic spiral patterns, while others become relatively featureless "elliptical" (actually ellipsoidal) galaxies (Figure 51)?

Figure 51: A typical elliptical galaxy, M49, in the constellation Virgo.
National Optical Astronomy Observatories

These are important questions, and the images of galaxies obtained with the Space Telescope will go a long way to helping astronomers to resolve them. But even more important results—because they offer the potential to resolve even greater mysteries—may well flow from the Space Telescope's extragalactic observations. We may learn just what produces the mysterious quasars.

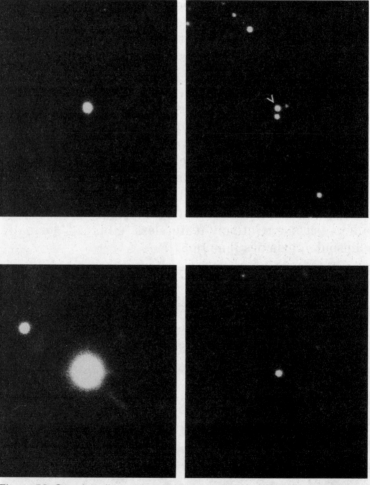

Figure 52: Quasi-stellar radio sources (quasars) appear to be points of light like stars, but are actually huge systems at vast distances in the universe. The quasar 3C 273 *(lower left)* has a jet of matter about 200,000 light-years long. *Palomar Observatory*

Quasars (an acronym for "quasi-stellar radio sources") are pointlike objects that look like stars—that is, they are "quasi-stellar"—on even the best photographs (Figure 52). Their starlike images prove that the energy from the quasars must originate in a relatively small region—far smaller than galaxies—because even the most distant galaxy is large enough to appear fuzzy rather than pointlike when photographed with a powerful telescope. The "radio source" portion of quasars' name signifies that some quasars, unlike most galaxies, are powerful sources of radio emission, which was what originally caught astronomers' attention.

But now that more than a thousand quasars have been found, the real news about quasars turns out to be this: Quasars are the most distant objects known and the most powerful sources of energy that we have found in the universe. Yet quasars' relatively small sizes prove that these incredible outpourings of energy arise in a region no larger than the innermost one ten-thousandth of the Milky Way, perhaps even from regions only one ten-thousandth of that size! In other words, for the past two decades, quasars have posed the conundrum of objects that produce far more energy each second than even a giant galaxy does, yet occupy a volume far less than one-trillionth the volume of the Milky Way. How can such vast amounts of energy arise within such tiny regions of space?

The current best hypothesis to explain quasars—"best" because it has not yet been contradicted by any stubborn observational facts—makes each quasar the site of a supermassive black hole, a black hole with as much as a billion times the mass of our sun. Black holes, first hypothesized scientifically by J. Robert Oppenheimer, are objects with such enormous gravitational forces that nothing—not even light—can escape from them.

More precisely, nothing can escape from within a certain distance, the object's "black-hole radius," from the center of the object. If our sun were compressed to a radius of two miles, instead of its actual 400,000 miles, the sun would become a black hole, and nothing could escape once it came within two miles of the sun's center. If you take an object with a million times the sun's mass, its black hole radius will be not two miles but two million miles. Squeeze the mass so that it all lies within two million miles of the center, and you will have a supermassive black hole—a black hole with a million times the sun's mass and an enormous gravitational attraction.

Two million miles sounds like an enormous distance, but not in terms of the vast distances within galaxies. Our own sun lies ninety-three million miles from Earth, and the solar system spans four billion miles or so. Hence a supermassive black hole with a million times the sun's mass could fit comfortably inside the Earth's orbit—inside all the planets' orbits—occupying only the tiniest fraction of the solar system, which is part of a galaxy a billion times larger than the solar system itself. Even a supermassive black hole with 100 million solar masses would have a black hole radius of "only" 200 million miles, a little larger than the orbit of the planet Mars. In comparison to an entire galaxy, this is still incredibly small.

Now picture such a supermassive black hole at the center of a newly forming galaxy, a vast cloud of gas and dust. The black hole's immense gravitational attraction tends to pull the gas and dust toward it. Because the cloud of gas and dust has some angular momentum of its own, it does not fall straight toward the black hole, for the same reason that an artificial satellite of the Earth—like the Space Telescope itself—does not fall to Earth but orbits around it. However, if any friction acts on the gas orbiting the black hole, it will spiral slowly inward, for the same reason that air resistance causes the Space Telescope to spiral—we hope very slowly—back to Earth.

As the gas spirals in, and before it comes inside the black-hole radius, it grows immensely hot as each of its constituent particles collides countless times with other particles, and at higher and higher velocities as the gas moves more and more rapidly along ever-tightening spirals. The high-temperature gas emits great floods of electromagnetic radiation, because any object not at absolute zero emits such radiation, and it emits more and more the higher and higher its temperature rises.

Thus, although no radiation—or anything else—can escape from inside the black-hole radius, from somewhat *outside* the black-hole radius we should see enormous amounts of such radiation, whose brightness masks the black hole within. Here we could have the explanation of quasars, each of them "powered" by a supermassive black hole that drags matter into ever-tighter spirals, heating it to such enormous temperatures that the matter's last gasp, before falling all the way in, produces extremely compact and powerful sources of radiation that we can see even at distances of ten or fifteen billion light-years.

How can we find out more about this with the Space Telescope? We cannot hope to observe a black hole, which by definition is invisible. But we can hope to see whether the hypothesis of supermassive black holes can survive a far more accurate check than has yet been imposed. The Space Telescope should be able to see whether, as hypothesized, a quasar consists of an extremely bright pointlike source, surrounded by more diffuse matter that will slowly spiral into the black hole. Astronomers who favor the supermassive-black-hole model of quasars expect to see gas around the quasar itself. This gas could be the "fuel" for the black hole. If, however, the Space Telescope, with its far sharper view than any Earthbound telescope, reveals no such gas—if, for example, we still see only a point of light and nothing more—then the theorists will have to return to their drawing boards.

In addition to its observations of quasars, the Space Telescope will also observe what are believed to be related objects, Seyfert galaxies and galaxies with active nuclei (Figure 53). These galaxies are thought to be something like quasars but perhaps on a smaller scale, with a black hole having "only" 100 million solar masses, not a billion or more as in quasars.

Seyfert galaxies and galaxies with active nuclei contain

Figure 53: NGC 4151, a Seyfert galaxy with an active, bright nucleus.
Palomar Observatory

central regions having intense energy outputs. Thus these classes of objects correspond more or less to the model that exists for quasars, since they too are supposed to contain supermassive black holes at their centers. However, Seyfert galaxies and galaxies with active nuclei are relatively pale copies of quasars in energy terms. They are not the most intense sources of energy that we know, but instead, as implied by the term *active nuclei*, are basically galaxies in which the central regions are particularly active.

Once again, by obtaining a far clearer view of the galaxies than heretofore possible, the Space Telescope should reveal details of these objects' structure that will help to explain how gas moves within them, presumably falling, from time to time, into the nucleus, where a supermassive black hole is believed to lurk.

Seyfert galaxies are far more numerous than are quasars, so the nearest ones are closer to us. Therefore, the chance that the Space Telescope will discern details of this process is correspondingly greater than for quasars.

While we are considering the subject of supermassive

Figure 54: M31, the Andromeda Galaxy, is the nearest large galaxy, having a distance of two million light-years. Note the two small elliptical galaxies accompanying it. Some of the stars seen are foreground objects in our own galaxy. *Lick Observatory*

black holes, we ought not to overlook a belief among astronomers that every large spiral galaxy, our own Milky Way included, may have a supermassive black hole at its center. Such supermassive black holes might have only a few millions of times the sun's mass, rather than the billions of solar masses hypothesized for quasars. Nevertheless, the thought that our galaxy and nearby spirals such as the Andromeda Galaxy (Figure 54) may have such supermassive black holes within them opens a fascinating new vista on black-hole possibilities.

Sadly enough, the Space Telescope cannot hope to obtain a good look at the center of our galaxy; there is simply too much obscuration by the interstellar dust particles that concentrate heavily toward the central plane of the Milky Way. Since our solar system and the galactic center both lie in this central plane, a huge amount of dust blocks our view of the center. But when we look at other galaxies, such as the great spiral in Andromeda, a happier result ensues. We see these galaxies from a more favorable angle, so we can study their central regions relatively unhindered by the absorption produced by interstellar dust.

Thus once again the Space Telescope offers a fascinating opportunity. We shall be able to study the innermost parts of the Andromeda Galaxy, and of other nearby galaxies, so well that we may be able to detect the gas flowing in toward the central, supermassive black hole—if it is there.

In all these areas of research—planetary astronomy, stellar evolution, interstellar matter, and extragalactic astronomy—the Space Telescope will open our eyes to a better understanding of the cosmos. But the greatest advances of all that flow from the Space Telescope are likely to occur in the largest sphere of all—cosmology, the study of the structure and evolution of the universe as an organic whole.

8

COSMOLOGY AND THE SPACE TELESCOPE

WHAT IS THE universe? When and how did it come into being? What is its structure? What is its ultimate fate?

Such questions have engaged human minds since ancient times and still animate arguments much as they did in ancient Greece. Each age has provided answers to these questions, often dependent on the religious and philosophical outlook of the times.

Greek philosophers stated that the planets, the moon, and the sun move against a firmament of stars fixed on an immense crystalline sphere. Not until the Renaissance was this notion of the sphere of fixed stars seriously questioned in Europe. In 1543, Nicolaus Copernicus proposed that the Earth orbits the sun, a concept that made our Earth just another planet. Copernicus's radical notions took many decades to win widespread acceptance, but they finally did. In the late eighteenth century, the British astronomer William Herschel conceived of the cosmos as a giant—but finite—system of stars, arranged in a flywheel pattern, and noted that the sun is just an ordinary star, far brighter to us than the others only because it is so close.

In the twentieth century, the Missouri-born astronomer Harlow Shapley found that the center of Herschel's cosmic system, which is now called the Milky Way Galaxy, lies in the direction of the constellation Sagittarius and has the enormous distance from us of 30,000 light-years. (A light-year, the distance traveled by light during one year, is

63,000 times greater than the distance from the Earth to the sun.) During the 1930s, the Dutch astronomer Jan Oort demonstrated that the Milky Way rotates around its distant center in about 250 million years.

From Copernicus to Shapley, our concept of the universe had expanded from the solar system to a galaxy a billion times larger.

GALAXIES

While astronomers were struggling to digest this new information, another fact had emerged. Working at the Mount Wilson Observatory near Pasadena, California, during the 1910s and 1920s, Edwin Hubble investigated the nature of certain fuzzy patches of light, originally cataloged by Charles Messier in the late eighteenth century. Were Messier's objects clouds of gas *within* our own Milky Way, as had long been believed? Hubble probed these objects with the aid of the famous 100-inch reflecting telescope built on Mount Wilson in 1917 (Figure 55). He found that although some of Messier's objects consist of hot gas, others lie far outside our own galaxy.

In 1924, Hubble found that Messier's object 31—M31 for short—has a distance from us of two million light-years, nearly 100 times greater than the distance to the center of our galaxy. (We give the modern distance, corrected for effects unknown to Hubble.) M31 is a giant star system like the Milky Way that contains hundreds of billions of stars; like our own Milky Way, this galaxy rotates every few hundred million years. M31, also called the Andromeda Galaxy, has spiral arms traced out by particularly bright stars that are accompanied by dust in interstellar space (see Figure 54).

Similar structure was later discovered in our own galaxy, and M31 seems to be truly a sister galaxy to the one in which we live. Both are giant spiral galaxies, characterized by a flattened disklike distribution of stars, within which bright stars and dust outline the galaxy's spiral arms.

Since Hubble's time, astronomers have examined and

Figure 55: The 100-inch telescope at the Mount Wilson Observatory, with which Hubble estimated the distances to galaxies, revealing the expansion of the universe. *Mount Wilson and Las Campanas Observatories*

Edwin Hubble at the sixty-inch Schmidt telescope on Palomar Mountain. *Palomar Observatory*

cataloged many thousands of galaxies. Some galaxies differ from M31 in having little or no flattening and no spiral arms, but only a smooth distribution of stars and spherical or ellipsoidal shapes. These are the elliptical galaxies (see Figure 51). Spiral and elliptical galaxies are remarkably similar to one another in their sizes and masses, and form the building blocks of the universe that we know today.

Will we eventually find that galaxies, like the stars forming our own galaxy, represent just a tiny agglomeration in a much vaster system, one much larger than the distance to the most distant galaxy? We can't provide the answer to this question today, but we do know that on the largest scales that we can observe, groups of galaxies move in a way that differs dramatically from the motions of the stars within the galaxies they form. The stars in each galaxy orbit around its center, but the *system* of groups of galaxies is expanding!

In 1912, Vesto Slipher, an astronomer at the Lowell Observatory in Arizona, first succeeded in measuring the Doppler shift in the spectrum of M31. The Doppler shift describes the fact that if a source of light moves toward or away from an observer, then all the wavelengths of light measured by an observer, and thus the entire visible-light spectrum of a star or galaxy, will appear to change. If the object is approaching the observer, the wavelengths will show a "blueshift," a change to shorter wavelengths. In contrast, if the object is receding, the entire spectrum will shift toward longer (red) wavelengths. In this process, called the Doppler effect, all the wavelengths change by the same fraction of their original value. The amount of this fractional change, called the Doppler shift, depends on the velocity of the object with respect to the observer.

The Doppler effect is familiar to us in everyday life, since it affects not only light but also sound waves coming from moving objects such as speeding ambulances. The sound of an approaching siren shifts in pitch from high to low as the ambulance first approaches and then recedes from us.

Christian Doppler, an Austrian physicist, gave the correct explanation of this effect in 1842 on the basis of the wave theory of sound. Doppler predicted that a similar shift should be found for moving objects emitting light waves, but the velocity of light is so enormous that any familiar velocity—say 100 miles per hour—is essentially zero in comparison, making the Doppler shift of light waves very difficult to detect in the laboratory.

The Doppler effect predicts that if a luminous object is moving away from the observer with a velocity v, all wavelengths of its light are shifted to longer wavelengths by the same fraction, equal to the ratio of v to the speed of light, c (which is 186,000 miles per second). If, on the other hand, the object moves toward the observer, the wavelengths of its light are shifted to shorter wavelengths by the same fractional amount, $v \div c$. Slipher observed that the wavelengths of the features in the spectrum of M31 are all shifted toward the blue (shorter wavelengths) by 0.09 percent. This allows us to infer that M31 is approaching us with a velocity equal to 0.09 percent of the speed of light, or 167 miles per second.

THE EXPANDING UNIVERSE

During the late 1920s, Edwin Hubble, who was the world's expert at estimating the distances to galaxies, seized on Slipher's results and began, with Milton Humason, to assemble the velocities of all the galaxies he could for comparison with their distances. Hubble found that the galaxies whose distances from us are of the same order as that of M31 (two million light-years) form a group, now called the Local Group. The galaxies in the Local Group have velocities that lie between −210 miles per second (approaching) and +230 miles per second (receding). Nowadays, we interpret these velocities as being the result of the orbital motion of individual galaxies, and conclude that the galaxies in the Local Group are moving under the influence of their mutual gravitational attraction around the center of mass of the entire Local Group, just as stars move in an individual galaxy.

But when Hubble and Humason turned their attention to more distant galaxies, they found *no* galaxies approaching us. All the distant galaxies outside the Local Group are receding from us! Furthermore, Hubble found that the greater a galaxy's distance from us, the greater is its velocity of recession. In 1929, Hubble announced his discovery, now called Hubble's law: for any galaxy outside the Local Group, the recession velocity v equals a constant (now called H, for Hubble) times the galaxy's distance, d.

Measurements of thousands of galaxies since 1929 have validated this relationship, $v = H \times d$, for galaxies out to distances approaching ten billion light-years, a hundred times the distances Hubble measured, and to velocities up to 100,000 miles per second—more than half the speed of light. If (and this is a big if) our view of the cosmos is representative, then the universe is expanding. All distant galaxies are receding from other galaxies at speeds proportional to their distances from each other.

We should add a note of clarification to the expansion phenomenon. Just as in the Local Group, almost all galaxies belong to one group of galaxies or another that contains a dozen or so galaxies, or to clusters of galaxies containing up to a thousand or more member galaxies. The expansion of the universe involves the recession of *groups* and *clusters* of galaxies. Superimposed on the overall recession are random individual motions of each galaxy, arising from its orbital motion within its cluster, typically at several hundred miles per second.

How should we understand the expansion of the universe? Does it arise from an explosion in our part of space, flinging the galaxies out into a much larger region? One might suppose so, because in that case, Hubble's law is just what we would expect. Suppose that the universe had a single, unique center, from which matter exploded at a time t in the past. Now, a time t later, each galaxy would have moved away from the center by a distance $d = v \times t$, where

v is its velocity. But this set of motions precisely follows Hubble's law, if we equate the Hubble constant, H, with 1 divided by t, t being the time since the explosion. We then have $v = d \div t$, analogous to Hubble's $v = H \times d$. Since H equals about fifteen miles per second of velocity for each million light-years of distance, if $1 \div t$ equals this number, then t would equal a million light-years divided by fifteen miles per second, which works out to be thirteen billion years. Could an explosion from one central point thirteen billion years ago have flung the galaxies into space at speeds up to half the speed of light? We could then understand why the most rapidly moving galaxies are also the most distant from us: these galaxies reached the greatest distances from us in the time since the explosion.

Although this simple model is appealing, it is wrong—at least according to astronomers. The model posits a unique center for the explosion, as well as an edge formed by the most rapidly moving galaxies. Unless we happened to be exactly at the center of the explosion—an extremely unlikely occurrence in view of the fact that our galaxy is no different from any other—we would observe more galaxies when we look in the direction toward the center and fewer in the direction away from the center. No such effect is observed in any direction. Instead, the number of galaxies, so far as we can tell, is the same in all directions.

MODERN CONCEPTS OF THE UNIVERSE

In contrast to the model of the universe in which the expansion centers on ourselves, consider the model that astronomers now believe to be correct. This model rests on what astronomers often call the "Copernican principle," in honor of the man who insisted that the Earth is not the center of the universe. According to the Copernican principle—which is only a working hypothesis—our view of the universe is a representative one; every observer in the universe sees just about what we see. Galaxy clusters are receding from *that* observer, with speeds that are proportional to the clusters' distances from *that* observer. In other words, galaxy clusters are moving away from each other everywhere.

If this is so, then space itself must be expanding, carrying the galaxy clusters along with it. If space simply "sat there," and galaxies expanded through it, then the expansion would have a real center, and we could find it by locating the point that is stationary in space. This would violate the Copernican principle, a principle of cosmic democracy, in which all observers are equal.

Although we cannot prove that the Copernican principle must be valid, astronomers insist that we should use it until it is disproved, for it offers by far the simplest possible view of the universe. Anything else would require that one or more points in the universe be special and different from all the rest—not an impossibility, but considered tremendously unlikely if we adopt the spirit of Copernicus. And if all observers see galaxy clusters receding from them, then no point can be stationary in space. Instead, all of space must be expanding like the surface of an expanding balloon, on which all points see all other points receding with speeds proportional to their distances (Figure 56).

The astronomically accepted model for the expansion of the universe makes all of three-dimensional space analogous to the expanding two-dimensional surface of a balloon. Simply imagine that you are a flat, nonexpanding dot on that surface, and that no inside or outside of the balloon exists. (Not so easy to do, but give it a try.) Then you will see that all points are equal, and that every dot must see Hubble's law hold true for that observer as space (the surface) expands.

Try to ignore the difficulties with the model, which, after all, is only an analogy: What does real space expand into? (Actually, it needs nothing to expand into.) What corresponds to the inside and the outside of the balloon? (Nothing at all, not even space, though you can use extra mathematical dimensions to model this imaginary "inside" and "outside.") Don't be discouraged! Your difficulties in con-

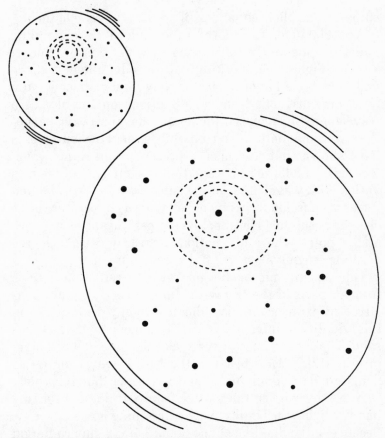

Figure 56: The two-dimensional surface of an expanding balloon can be used to model three-dimensional space in the expanding universe.
Drawing by Marjorie Baird Garlin

ceiving of all of three-dimensional space as expanding, everywhere, are shared by millions—or at least they would be if cosmological theories were more widely discussed.

Remember that human intuition was formed in a tremendously limited part of the universe, our little corner of space. It's hardly surprising, then, that when we try to imagine the entire universe, we can't step back, mentally or physically, to obtain a clear view, like our view of an expanding balloon! We must try to make do with analogy— and with what cosmologists have learned from their hypotheses and their equations, if we care to believe them.

A crucial item of support for the model of the expanding universe appeared in 1965, when Arno Penzias and Robert Wilson (Figure 57), working with a radio antenna at the Bell Telephone Laboratories in New Jersey, discovered a faint hiss arriving from all directions on the sky at a wavelength of seven centimeters in the microwave part of the electromagnetic spectrum. Subsequent measurements have shown that this hiss is present at all radio wavelengths over a large range and that the intensity of the hiss varies with wavelength in the same way as the radio and infrared emission from any object whose temperature is 2.8 degrees absolute. (The absolute temperature scale has zero at the point where all motion ceases, and each degree absolute is equivalent to 1.8 degrees Fahrenheit.)

This cosmic microwave background radiation arrives from all parts of the sky with almost exactly equal intensity. Recent observations indicate that any deviation from this equality of intensity is less than one part in ten thousand. The cosmic microwave background cannot originate in our Milky Way, because the Milky Way's matter is distributed in a giant disk 100 times wider than it is thick; that is why we see the matter from within as a narrow band of light (the "milky way") stretching across the sky. Because of the disklike shape of our galaxy, any radiation from the Milky Way would arrive from a band across the sky, completely unlike the observed cosmic microwave background. Instead, the cosmic background radiation has a distribution over the sky that is uniform to better than one part in ten thousand.

This fact rules out a naive model of the expanding universe that we discussed earlier. If the cosmic microwave background arises from the universe as a whole, then in order to explain its uniformity, our galaxy would have to lie at the center of the explosion to within one part in ten thousand of the size of the universe. One can calculate the probability of any galaxy taken at random lying so close to the center to be less than one in a trillion. Thus, unless we

Figure 57: *(Left)* Arno Penzias (1933–   ) and *(right)* Robert Wilson (1936–   ), who first detected the cosmic microwave background radiation with the radio telescope seen behind them. *Bell Laboratories*

assume that a special providence has placed us at the center of the universe, we must look for another, better interpretation of the expansion we observe.

THE BIG-BANG MODEL OF THE UNIVERSE

This better interpretation, now called the big-bang model, was proposed by Alexander Friedmann, a Russian meteorologist and mathematician, in 1922, seven years before Hubble announced his discovery of the law of the recession of the galaxies. Friedmann's model is based on Albert Einstein's general theory of relativity, a generalization of Newton's theory of gravitation that Einstein proposed in 1915, and which has been successfully tested many times since.

According to Einstein's general theory, gravitational

force may be thought of not as a conventional force but as curvature of the space-time in which we live. In this interpretation, gravity bends space. For example, the sun's gravity bends nearby space—by greater amounts closer to the sun—and nearby objects respond to this bending by deviating from what would otherwise be straight-line motion. Thus orbital motion like the Earth's motion around the sun reflects the response of celestial objects to the curvature of space-time. General relativity gives the same results as Newton's theory of gravitation when all velocities involved in the motion are much smaller than the speed of light, as is true within our galaxy. But in the expanding universe, where the velocities of the most distant galaxies are observed to be half the speed of light or greater, we would expect major deviations from Newton's theory. This gives us a chance to test Einstein's theory of relativity.

OPEN, CLOSED, AND FLAT UNIVERSES

In general relativity, the amount of curvature of space-time of a given region depends on the amount of matter within that region. Friedmann wondered if Einstein's equations had a consistent solution that would describe the entire universe, supposing that matter is distributed uniformly throughout the universe. (We now know this to be approximately true when we average over regions as large as 100 million light-years on a side.)

Remarkably, Friedmann found three exact solutions to Einstein's equations, which are called open, closed, and flat models of the universe. In each solution, the amount of curvature of space-time at any given time is the same everywhere throughout the universe. But the amount of curvature does change with time; it always starts with an infinite amount of curvature at a time we will call $t = 0$, and then decreases as the years go by. The time $t = 0$ marks the moment of the big bang or the "initial singularity" that characterizes all three of Friedmann's models.

At time $t = 0$, all of the universe—all of space too!—occupied far less room than it does now. At that time, ten to fifteen billion years ago, all the matter in the universe was compressed to (theoretically) infinitely high temperatures and infinitely high densities. Since infinities make physicists nervous, the Friedmann models technically begin at some tiny moment—as tiny as you can choose—after the initial singularity. But for our purposes, we can round off, and say that the big bang marks the moment when the universe began its present expansion, with a density and temperature of matter higher than any number you care to name.

Ever since the big bang, the universe has expanded. This cosmic expansion has lowered the density of matter in the universe, since the same amount of matter (according to the standard theory) has filled a progressively greater volume. The expansion has also lowered the temperature, because matter expanding into a larger volume cools off like the spray escaping from a pressurized can, which feels noticeably cooler than the surrounding air. And just as the density and temperature have fallen since the big bang, so has the curvature of space-time decreased.

The key difference among the three Friedmann models resides in what will happen to the curvature in the future. In the open model, the curvature decreases forever. In the closed model, the curvature decreases to a minimum value, after which time it increases again in just the reverse process of its decrease, and it eventually reaches infinity once again. In the flat model, the curvature decreases forever in a special way, falling to zero only after an infinite amount of time has passed (Figure 58).

In each of these three models, the curvature of space that corresponds to the curvature of space-time is different. In the open model, space curves outward, with ever-increasing amounts of space at large distances. In the closed model, space curves in upon itself, with smaller amounts of space at large distances. In the flat model, as its name implies, space does not curve at all. The closed model contains a finite amount of space, but in both the open model and the flat model, an infinite amount of space exists.

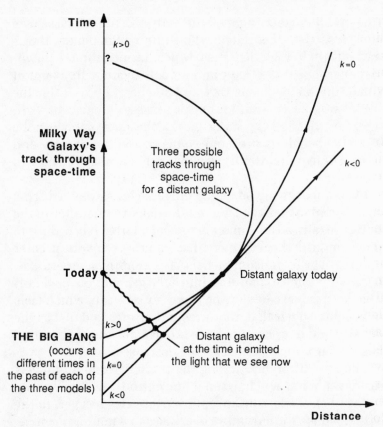

Figure 58: In the three types of Friedmann models of the expanding universe, the distances between galaxies vary with time. The value of *k* indicates the curvature; *k* = 0 is the flat-space model, *k* < 0 is the open model, and *k* > 0 is the closed model. *Drawing by Marjorie Baird Garlin*

WHAT IS CURVATURE?

The curvature of space presents a challenging concept, but one that is possible to understand. The best way to envision this curvature, at least for the closed model, is to consider the surface of the Earth or, better yet, a terrestrial globe. We must, however, attempt to believe that the globe's surface represents *all* the space in the universe. If we close our eyes and think only of the surface itself, we can perhaps accept a two-dimensional analogy to three-dimensional space. We can move on the surface in only two independent directions: north-south or east-west. If we move up or down, we move off the surface of the globe, which does not help us to use the analogy.

We spoke of a closed universe as one in which space curves in upon itself. In that case, we were considering three-dimensional space, but this concept also applies to a globe. The globe has a finite surface area, not an infinite amount, as would be true for a flat plane extending to infinity in all directions. On the surface of a globe, "straight lines" curve because the surface curves. Still, there are lines that are as close to being straight as is possible, called great circles. A great circle through any point is the largest circle that can be drawn through that point, formed by imagining a plane that passes through both the point on the surface and the center of the globe. The points where the plane intersects the globe form the "great circle." On the Earth, great circles are important for air transportation because they represent the shortest routes between any two points; that's why you find yourself passing over Greenland when you fly from Los Angeles to Paris. In the sense of providing the shortest routes, great circles on a globe resemble straight lines on a flat surface. Even though great circles are curved—because they are drawn on a curved surface—they are as close to being straight as possible, because you do not deviate left or right as you move along a great circle.

Straight lines on a flat surface extend forever; a traveler on a flat surface who moves along a straight line never returns to the place where he or she started. On a globe, however, a traveler on a great circle does return to the point of origin after completing one circumference of the globe. Although they took a zigzag route, that is exactly what Ferdinand Magellan's crew did from 1519 to 1522, when they circumnavigated the Earth for the first time.

This is what we mean when we say that the surface of a globe "curves in upon itself": lines that are as straight as possible return to themselves. This is quite different from a flat surface, on which such lines extend without limit. Note, however, that travelers on the globe never come to an edge; no chasm or wall exists to impede their progress at any point. Before Columbus and Magellan, some sailors fretted about falling off the edge of the Earth, probably because they could not believe that the Earth is flat yet infinite. They rejected the possibility that the surface of the Earth is finite and yet has no edge—which is the simple truth about its two-dimensional surface.

Note also that the surface of the globe has no center. "What?" you say. Of course the Earth does—it's deep in the Earth, halfway between points opposite each other on the globe. Not fair! Remember that you must imagine a truly two-dimensional surface, that of a globe, existing independently from the interior of the globe or the space that we know exists outside it. There is no center for the surface itself, and it is the surface that is representing three-dimensional space. To see why, imagine that some city nominates itself to be the Center of the World. Although Tokyo, Mexico City, and New York could claim that title on the basis of their population, it wouldn't stick. Geographically, none of these cities has any more special position than any other or, indeed, than any position whatsoever on the surface of the globe.

To summarize, the two-dimensional surface of a globe has no center and no edge but is nevertheless finite because it curves back upon itself. Now try this feat of imagination: Try to imagine a three-dimensional space having these properties—the lack of any center or edge—yet still finite in total extent. If you succeed, congratulations! Most people, including the authors, can't. But this is the type of space that we are talking about when we speak of a closed universe. Literally, if you could travel fast enough in a "straight line" in such a universe, you would sooner or later return to your starting place.

Why is it so difficult for us to imagine a closed universe? Probably because the space in and around our galaxy is so nearly flat that we have never experienced curved space directly. Furthermore, we can envision curvature only when we have an extra dimension to see space curving into. This method works for imagining the two-dimensional surface of a globe, but not for all of three-dimensional space.

However, we should be careful not to conclude from our intuition that space must actually be flat. To do so would represent a mistake in reasoning like that of the ancient philosophers who concluded that the Earth is flat. Since everyday experience fits the flat-Earth concept, they concluded that *all* of the Earth's surface must be flat. We must attempt to avoid such mistakes if we can. We should admit that we have experienced only a tiny portion of the space in the universe, and that space could be curved, so that the entire universe could consist of closed space, like the surface of a sphere.

DETERMINING THE CURVATURE OF SPACE

Remember that we are considering Friedmann's *models* of the universe, not the universe itself. As scientists, we must make observations to decide whether or not any of Friedmann's models describe the real universe and, if so, which model—open, closed, or flat—fits the data best.

As its name suggests, ordinary Euclidean geometry applies to lines drawn in the flat model, which is infinite in extent. The open model is also infinite, but Euclidean geometry does not apply to it. We can summarize the geometrical differences among the models by considering the sum of the angles in a triangle whose sides are as straight as they can be, though still constrained by the underlying curvature of space (Figure 59). In the flat universe, the sum of the angles equals 180 degrees, as in Euclidean geometry. In a closed universe, the sum exceeds 180 degrees; in an open universe, the sum is less than 180 degrees.

To see why the sum of the angles exceeds 180 degrees in

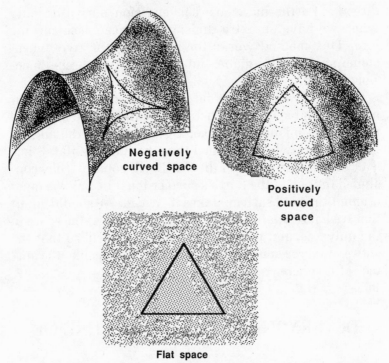

Flat space

Figure 59: The three types of space can be distinguished by whether the sum of the angles in a triangle is 180 degrees (flat space, at bottom), less than 180 degrees (negatively curved space, upper left), or greater than 180 degrees (positively curved space, upper right). *Drawing by Marjorie Baird Garlin*

a closed universe, consider a triangle made of portions of three great circles on a globe; these lines are as straight as they can be, given the curvature of the surface. Let us choose the three sides as (1) the equator; (2) a north-south meridian at longitude zero degrees, through London; and (3) a north-south meridian at longitude ninety degrees, through St. Louis. Since both meridians are perpendicular to the equator, the angles where they intersect the equator are each ninety degrees. The meridians meet at the North Pole, where, because of the longitude differences, the angle between them also equals ninety degrees. Therefore, the sum of the angles is $90 + 90 + 90 = 270$ degrees, a good deal

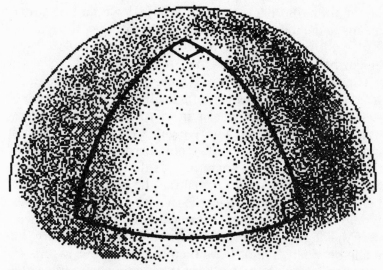

Figure 60: A depiction of the surface of the Earth, showing how the three angles in a triangle that consists of portions of great circles add up to 270 degrees. *Drawing by Marjorie Baird Garlin*

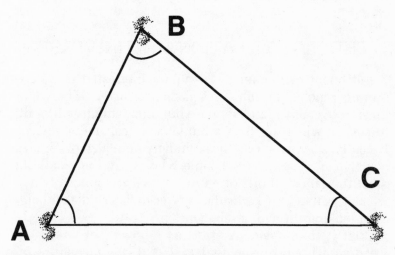

Figure 61: Observers on three different galaxies try to determine the curvature of space by measuring angles A, B, and C in a hypothetical experiment. *Drawing by Marjorie Baird Garlin*

larger than 180 degrees (Figure 60). If we had no other
means to demonstrate the positive curvature of the Earth's
surface, it would suffice to show that this triangle's three
angles have a sum greater than 180 degrees.

How can we ever find out which, if any, of Friedmann's
three models is correct? Hypothetically we could try to find
the sum of the three angles of an enormous triangle (Figure
61). Find friends in each of two galaxies, B and C, remote
from each other and remote from ourselves, located at A
(naturally). At a given instant, say midnight on December
31, 1999, each observer measures the angles on the sky
between the galaxies where the two friends are located. We
would measure the angle between B and C; B would measure the angle between A and C; and C would measure the
angle between A and B. We then broadcast our results to
our friends, and they broadcast to us, using powerful radio
transmitters. After waiting for millions of years (because
the distances involved are millions of light-years), we add
the three observed angles in order to decide which model
describes the universe we live in. The practical problems
with this theoretically correct method are obvious.

THE NEW INFLATIONARY UNIVERSE

Astronomers remain passionately interested in determining which of Friedmann's models is correct. The problem is extremely simple to pose but maddeningly difficult
to solve—which in itself stimulates astronomers' juices.
Recently, scientists from the seemingly unrelated discipline
of particle physics—the people who study the smallest,
most fundamental bits of nature—have announced to astronomers that they know the answer to this riddle without
even looking through a telescope!

Briefly, the physicists argue as follows. Let us accept
Friedmann's conclusion that at $t = 0$, the curvature of
space-time was infinite, that is, matter was then infinitely
compressed. Taking a cue from the observation of the

cosmic microwave background, we may persuasively argue that the universe must have been hotter than 2.8 degrees absolute long ago, because it was compressed; in fact, the universe should have been infinitely hot at time $t = 0$. We then turn to modern particle-physics theories, particularly to what physicists call the Grand Unified Theories (GUTs), and note that something strange must have happened at $t = 10^{-35}$ second (ten billionths of a billionth of a billionth of a billionth of a second).

At that time, an instant after the big bang, the temperature throughout the universe was 10^{28} degrees (ten billion billion billion degrees), so hot (even for particle physicists) that all the forms of matter that we know today—plus any that we still may not have encountered in particle accelerators—were converted into a kind of primitive matter that we can never hope to create, even with our greatest expenditures of money and effort. As the universe expanded, this primitive material began to cool.

A strange thing happened—so the particle physicists tell us—at about 10^{-35} second after the big bang. Just as water freezes into ice when it cools sufficiently, the primitive material "froze" into stuff more like what we see around us today, such as matter made of electrons and quarks (the constituents of protons and neutrons).

This process is just what the physics doctors ordered to make the stars and planets that we find in the universe today. But the "freezing" of ordinary matter out of unfamiliar matter has an additional twist. Particle physicists calculate that, at the time that this freezing occurred, the expansion of the universe must have speeded up to truly enormous velocities, with material rushing apart—everywhere—so rapidly that distances between particles quickly became ten trillion trillion times what they had been before. This "inflation"—the expansion of the entire universe at enormous velocities—occurred because the process of freezing from one mix of particle types to another released energy.

When this period of inflation was over, a mere 10^{-32}

second after the big bang, the universal expansion slowed to its "normal" pace. We would hardly know that inflation had occurred except for the key fact that if inflation *did* occur, it must have forced the geometry of the universe to become flat, or so nearly flat that any deviation from perfect flatness is academic.

Why would inflation have flattened space in the universe? There is a simple analogy that might help: the real terrestrial globe. Transport yourself in your imagination to the middle of the ocean on a very calm day. "The sea is flat," you say, and you would be right, or nearly so. It would surely *look* flat, but you know it really isn't, because after all the world is round. The reason that the sea looks flat is that you observe only a tiny part of it. Even when you fly over the ocean in an airplane, your horizon includes at most one or two hundred miles, far less than the twenty-five-thousand-mile circumference of the Earth, and the sea looks flat from an airplane too. To get a true sense of the Earth's roundness, you must rise to the altitudes where the Space Shuttle orbits the Earth, several hundred miles up, where you can see for thousands of miles around the Earth. There it is obvious that the Earth is a globe. But if the Earth suddenly became a billion times larger than it is now, not even our best spacecraft could rise high enough to reveal its curvature.

Like the Earth, the universe as a whole may be curved, but inflation may have made the universe so large that the part we see forms only a tiny part of the whole and looks flat. We are talking about a hypothesized inflation that made the universe not a few thousand, or a few trillion, times larger, but 10^{25} times—one followed by twenty-five zeros—larger than it had been! Just as the sailor perceives the neighboring ocean as flat, we sailors in cosmic space perceive our part of the universe, everything that we can see, as flat. The particle physicists tell us, "The real universe is much, much bigger than you astronomers think, and it was made that way by inflation during the first 10^{-32}

second. Take it from us, when you are finally able to make accurate measurements, you will find that our piece of it will look flat."

Inflation also predicts that if the Hubble constant is 15 miles per second per million light years of distance, the average density of matter of all types in the universe should be about 10^{-29} times that of water. Astronomers have tried to test this prediction by calculating how much matter must be in each galaxy to account for the gravitation that holds it together. They find that the inner luminous parts of galaxies alone contribute about 1 percent of the predicted density. However, there is evidence that there is much more matter in the outer parts of galaxies, adding up to 10 percent of the predicted density.

Another experimental test is based on the fact that the relative abundances of isotopes of hydrogen, helium, and lithium produced by nuclear reactions in the big bang depend on the density of ordinary matter—that is, matter made of quarks and electrons. The predictions fit the observations of these isotopes well if the density of ordinary matter is between 4 percent and 6 percent of the predicted density of matter of all types.

It seems, therefore, that ordinary matter can account for the inner parts of galaxies and, conceivably, the outer parts as well but not the total density predicted by inflation. Either the theory predicting inflation is wrong or else there is "strange matter" in the universe—matter of some type not known on Earth. It would have to constitute more than 90 percent of the total. There is a tantalizing hint in the data that some such matter may reside in outer parts of galaxies.

The physicists believe that the universe must be flat and must contain the predicted density of matter. The astronomers regard it as an observational question. By allowing studies of galaxies and their interactions at great distances from us, the Space Telescope will help to test the inflationary model.

THE HORIZON OF THE OBSERVABLE UNIVERSE

We have referred several times to "the part of the universe that we can see." What does this mean, exactly? To astronomers, this phrase has a well-defined meaning, arising from the fact that, because of the finite speed of light, we now observe distant galaxies not as they are, but as they were. Looking out in space means looking back in time. This effect is negligible in everyday life, because light takes less than one hundred-thousandth of a second to travel a mile. But in the universe, distances are so large that the delay caused by light's travel time becomes substantial.

Since the expansion of the universe began some ten billion years ago, we might hope, by looking out far enough, to see the big bang itself. Certainly we can't hope to look out to even greater distances, because the universe didn't even exist at the corresponding earlier times! In effect, the observable universe has a "horizon," analogous to what we encounter on Earth, beyond which we can't see. However, the reason for the horizon differs for the two situations: For Earth, the horizon exists because the Earth is curved; for the universe, the horizon exists because the universe has existed for only a finite amount of time, and when we look out in space, we can't hope to observe anything that happened before that.

The horizon of the universe is a strange thing. If our observational instruments were powerful enough, we could penetrate right up to the horizon, ten to fifteen billion light-years away. There we would see infinitely dense, infinitely hot matter, just being born at the time of the big bang. However, because of the expansion of the universe, this matter would appear to be receding from us at the speed of light, so the radiation we might hope to see would be greatly attenuated.

Though we can conceive of this vision, a barrier separates us from the horizon, an impenetrable cauldron of hot gas, receding from us at 99.9999 percent the speed of light. This hot gas consists of the electrons and protons that cooled greatly since the time of inflation but are still seen at much hotter temperatures (about 6,000 degrees Fahrenheit) than those in the diffuse material in the universe today. This hot gas emits the radiation that we observe today as the cosmic microwave background, and we see it as it was about a million years after the big bang. Thus, observations of the cosmic background radiation take us 99.99 percent of the way back in time to the big bang. This is as far back in time as we can reasonably hope to observe with electromagnetic radiation.

Don't confuse the concept of the *observable* universe with the entire universe. Far beyond our horizon is material that should have reached conditions similar to those we see around us today. For all we know, intelligent beings are out beyond our horizon, puzzling about *their* horizon. Even in principle, we cannot observe that matter and those beings, because the light by which we would observe them would have had to have left before they came into existence, if it were to have reached us by now. The observable universe is simply that part of the universe that we can hope to observe, but there is far, far more to the universe out there—we think!

OBSERVATIONAL COSMOLOGY: GALAXIES IN FORMATION

What can astronomers hope to learn about the observable universe? We have already mentioned the cosmic microwave background radiation, which was emitted by hot gas from distances nearly at the observable horizon, receding nearly at the speed of light. Astronomers have been studying the characteristics of the background radiation over the last twenty-five years. Remarkably, the background appears to be featureless, at least within the accuracy of measurement (one part in ten thousand). From this,

we deduce that at the time the universe cooled to about
6,000 degrees Fahrenheit and emitted this radiation, the hot
gas was distributed throughout the universe extremely
smoothly.

That conclusion contrasts enormously with what we see
today. If we consider galaxies as building blocks of the
universe—ignoring the detailed structures inside them
such as stars and planets—we find enormous contrasts in
the density of matter from place to place. Within a galaxy,
the density of matter (the amount of matter per unit volume) is a million times greater than the density between
galaxies.

If we make an analogy with the steam rising from a pot
of hot water, gas in the early universe was distributed
uniformly, like the water vapor rising from the pot. But
today that same gas has formed galaxies—droplets like
those in the steam that we see higher above the pot. Condensation has occurred in which the smooth gas of long ago
produced the galaxies we see today.

Physicists tell us that galaxies must have formed
through condensation induced by gravity, in which minor
disturbances in the distribution of matter in the early
universe, although too small to affect the microwave background radiation, pulled themselves together through their
own gravitational forces to condense into galaxies at some
epoch during the past ten billion years. An entire branch of
astronomical research has grown up around this concept,
as astronomers use computers to simulate how the formation of galaxies may have occurred.

At present the most plausible idea seems to be that most
of the matter in the universe—at least 90 percent—consists
of particles of a type not yet detected in any laboratory.
These particles, called "cold dark matter," respond readily
to gravitational forces, and thus condense into objects of
galaxy-like dimensions. The ordinary matter with which
we are familiar (composed of quarks and electrons) would
also have been drawn into the condensations made of cold
dark matter, and would have condensed into stars to form
the galaxies that we see today.

Computer simulations suggest that this process began about half a billion years after the big bang, when the universe had 15 percent of its present size. (By "size" we refer to the distance between any two representative points, which increases as time passes and the universe expands.) Galaxy formation reached its peak of activity 1.3 billion years after the big bang, when the universe had 25 percent of its present size, and is nearly over today.

Astronomers are now trying to test these simulations against observations. This will be difficult to do, because in this model the peak of activity occurred early and therefore can be observed today only at a great distance from us. Hubble's simple rule allows us to estimate the speed at which a newly forming galaxy (as we see it) is receding from us. As the universe expands, the wavelengths of all types of radiation expand in proportion. Because the peak of galaxy formation occurred when the universe had a quarter of its present size, all wavelengths have increased four times since then. This corresponds to a speed of recession of a galaxy born at that time that equals 88 percent of the speed of light.

CANDIDATES FOR OBSERVATION

Astronomers have a problem in determining where to search for immensely distant galaxies in the process of formation. Figure 62 shows the image of a region of the sky that does not include any known stars or galaxies. The image was made with a sensitive detector at a large telescope, exposed for six hours—nearly all night long. All of the "twenty-sixth-magnitude" images, about as faint as any we can photograph from Earth, are believed to be distant galaxies. However, we have no idea, in view of the fuzziness of the images and the lack of spectral information about them, as to what their Doppler shifts, and hence their distances, may be.

Instead, we can examine a special class of galaxies that emit intense amounts of radio waves. Only a small fraction of all galaxies do this, and their radio emission is so strong

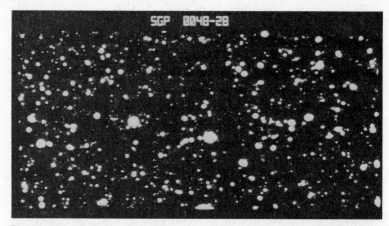

Figure 62: An image made using a new technique by J. A. Tyson and P. Seitzer at Cerro Tololo's four-meter telescope and a CCD camera, showing objects down to twenty-sixth magnitude. Most of these objects are believed to be very distant galaxies. *National Optical Astronomy Observatories*

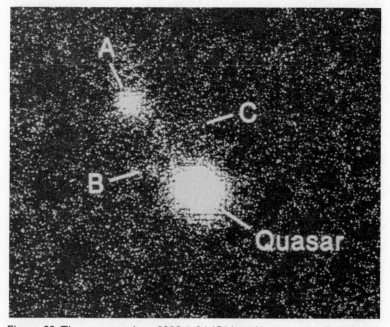

Figure 63: The young galaxy 0902 + 34 (Object A), seen at a distance of nine billion light-years, in the light of the Lyman-α line of hydrogen; it happens to be very close on the sky to a bright quasar. *Astrophysical Journal*

that we can detect them at great distances. We know this because in any random sample of radio sources we are likely to find several extremely distant galaxies. In this way, a dozen or so radio galaxies have been found with Doppler shifts indicating recession velocities greater than 70 percent of the speed of light.

Figure 63 is an image of an object called 0902 + 34, so far away that its light was emitted when the universe had only 23 percent of its current size and the age of the universe was only 1.1 billion years, compared to 10–15 billion years now. The image appears round, but it is rather extended on the sky. Its brightness falls off from its center into the background of light at a distance of about five arc seconds (equivalent to 150,000 light-years of actual size). This object appears to be much larger than galaxies near to us, which are about thirty thousand light-years in diameter. Perhaps this is not a single galaxy at all, but rather a small cluster of galaxies.

This object, and others like it, emit enormous amounts of electromagnetic radiation at a wavelength known as the Lyman-α (pronounced Lyman-alpha) line. This line is the most prominent spectral feature in the ultraviolet spectrum of hydrogen gas because hot hydrogen atoms naturally emit radiation at this wavelength. Although Lyman-α is prominent in the spectrum of hydrogen atoms observed in the laboratory and is known to be produced in copious quantities wherever hot luminous stars heat interstellar gas in their vicinity, the existence of this hydrogen emission from 0902 + 34 is something of a mystery. Why? When we study the gas near hot stars in our own galaxy, we observe very little Lyman-α radiation, even though the hot stars certainly produce it.

This result can be explained if the Lyman-α radiation is all absorbed by microscopic dust particles present in interstellar space close to the stars. Mixed in with the hot gas that produces the Lyman-α radiation, the dust particles block and absorb the radiation before it can propagate to us to be detected by our telescopes. Evidently this process does not occur in objects such as 0902 + 34, because we observe that their intense Lyman-α radiation amounts to as much as 100 million times the energy output of the sun. This implies that dust must be absent from such objects.

Another puzzle relates to whether the Lyman-α radiation from these objects really does arise from the heating of interstellar gas by hot stars. For most of the hot luminous stars in our galaxy, this heating process is accompanied by the stripping away from any carbon atoms of one or two of the atoms' outermost electrons, but no more; even the radiation from normal hot stars can strip only two electrons per atom. But in objects such as 0902 + 34, astronomers also observe carbon with *three* electrons missing. Since it takes more energy to do this than normal hot stars can produce, some other process must also be at work, of unknown origin.

Finally, 0902 + 34 emits greater amounts of Lyman-α radiation in comparison to other wavelengths of radiation (for example, hot stars themselves) than any other object yet observed, and the total amount of energy emitted as Lyman-α radiation equals 100 million times the energy output from the sun. All this makes astronomers suspect that 0902 + 34 must be quite different from any object we have ever observed.

THE SPACE TELESCOPE AS RESOLVER OF MYSTERIES

The Space Telescope is an ideal instrument for solving the mysteries of 0902 + 34 and other objects like it. Because the telescope will obtain images fifteen times sharper than those from ground-based telescopes, astronomers will be able to observe any details in 0902 + 34, such as groups of hot stars like those in other galaxies, with a clarity unattainable on Earth.

The Space Telescope will prove useful in another way— in studying the Lyman-α emission from objects, such as galaxies, that are much closer to us than the mysterious 0902 + 34. No other telescope can do that, because the

Lyman-α radiation never reaches the surface of the Earth. Unless the object has a tremendous recession velocity, its Lyman-α emission will be observed deep in the ultraviolet part of the spectrum. Therefore, our atmosphere blocks Lyman-α radiation from nearby galaxies, so ground-based telescopes cannot detect it. Astronomers can detect Lyman-α emission from 0902 + 34 only because this object is so distant that its Lyman-α radiation, along with all the rest of its emission, is shifted to much longer wavelengths because of the Doppler effect in the expanding universe. The Doppler effect shifts the Lyman-α emission into the middle of the visible part of the spectrum, where ground-based telescopes can readily detect it.

Orbiting above the ultraviolet-blocking atmosphere, the Space Telescope can observe Lyman-α radiation from any object, including those with nearly zero redshift, that is, objects quite close to us. Because we can study nearby objects, astronomers should be able to analyze in great detail how their Lyman-α radiation is produced. The information gleaned from such studies can be applied to distant objects such as 0902 + 34, and thus we can hope to learn their true nature.

While the Space Telescope observations are under way, theorists will produce detailed computer simulations of how galaxies should form in a universe dominated by cold dark matter. Comparison of the theoretical models with observations will tell us whether the models are viable or not. If there is an irreconcilable contradiction, we can conclude that cold dark matter is not the dominant component of the universe, contrary to the predictions of many particle physicists.

HOW TO TEST THEORIES

How do scientists claim to find out "the" truth? That is, which theories are held to be almost certainly correct, and which ones aren't? The goal is to find a crucial experiment, one for which opposing theories make different predictions of the experimental result. If we do, when the experiment is done, we may find that the predictions of theory A are contradicted and the predictions of theory B are fulfilled. If so, theory A must be judged incorrect, and that should be the end of theory A, which will be dropped from further discussion as scientists turn their attention to the task of testing theory B against other experiments.

It is important, however, to realize that even though theory A is actually false, it could be consistent with a particular experiment, since a particular experiment can only disprove a hypothesis, never provide complete confirmation. Furthermore, the proponents of theory A will continue to fight for it—as they should—until they become completely convinced that observations disprove it.

An example of this struggle among theories appears in the controversy that arose some years ago between the proponents of two radically different models of the universe. One model, the big-bang theory, states that the universe was formerly much denser than now, and has grown cooler and more rarefied. The other model, the "steady-state theory," proposed by Fred Hoyle, Hermann Bondi, and Thomas Gold, states that the universe maintains a steady state and never changes over any length of time. Although this model seems to contradict the observed expansion of the universe, its proponents hypothesized that new matter continuously appears from *nothing* throughout the depths of space. Thus, in the steady-state model, new galaxies continually form and maintain a constant average density of matter in the universe. The problem of creating matter out of nothing is a serious one, but the creation of matter also occurs in the big-bang model, at a single moment in time rather than continuing over billions of years.

For some time, astronomers had no way to choose between the two rival theories. Then, during the 1950s, radio astronomers began to discover radio sources so powerful that they could be observed even if they were located out toward the edge of the observable universe. The big-bang theory could make no specific predictions about how the numbers and characteristics of such radio sources should vary with their distances from us. However, since we observe the sources as they were in the past (because of the long travel time for radio waves) and since the overall properties of the universe have changed with time, the big-bang theory can accommodate the observation that the numbers of radio sources at large distances (hence the numbers existing long ago) differ markedly from the numbers of radio sources nearby (those that we observe in the astronomical present).

One virtue of the steady-state theory is that it could and did make a clear prediction about the numbers of radio sources. Because in the steady-state model the characteristics of the universe do not change with time, the numbers of radio sources must remain the same at all times in the past. Though radio sources at greater distances are observed as they were long ago, it was agreed that if the steady-state model is correct, astronomers should observe equal numbers of sources per million cubic light-years at all distances.

This prediction could be tested—and it was, by the British radio astronomer Martin Ryle and his co-workers. The results showed that the numbers of radio sources increase greatly at great distances, in sharp disagreement with the prediction of the steady-state theory. Scientists therefore concluded that the steady-state theory must be incorrect (or, as is sometimes said with more emphasis, dead). Today few theorists spend much time on the steady-state model, not because the model is inelegant or unappealing, but because observations have contradicted its predictions.

Since the big-bang theory is compatible with the observed different numbers of radio sources at large distances, the observations of radio sources do not contradict it. On the other hand, this does not mean that the big-bang theory is necessarily the correct theory. We simply have no proof that the theory is incorrect; it does correspond to what we have observed about the universe. We must remember, as the astronomer Martin Rees has said, that "absence of evidence is not evidence of absence." Other theories could prove equally compatible with the data, so we need additional tests.

Astronomers have now pushed their argument beyond the controversy over steady state versus big bang. The steady-state model is incorrect, but *which* big-bang model (if any) is correct? One Friedmann model or another (open, flat, or closed) is consistent with almost everything we observe: the fact that the universe is expanding, Hubble's law for that expansion, the numbers of radio sources at various distances, the cosmic microwave background radiation, and the abundances of the light elements, which were produced soon after the big bang.

But we are still not certain which Friedmann model fits the observations best. For example, if we assume that all matter in the universe is ordinary matter (that is, made of quarks and electrons), then the measured abundances of the light elements, which depend on the number of quarks in the early universe, indicate that the universe is open. But as we have explained earlier, evidence points to dark matter inside galaxies. If even larger amounts of dark matter exists outside galaxies, and if this dark matter is not ordinary but "strange" in the sense that it is not made up of the particles (quarks and electrons) that we find in ordinary matter, then the universe could be flat or even closed and still be consistent with the observed abundances of light elements. This is so because all the chemical elements consist of quarks and electrons, and the production of light elements in the early universe does not depend on what other particles were and are present.

BETTER VALUES OF THE HUBBLE CONSTANT: THE CEPHEID VARIABLES

The Space Telescope can help to resolve whether any Friedmann model fits the observations of the universe and, if so, which model. It can do so by studying objects at larger distances with unprecedented clarity. Since the three Friedmann models differ most in what they predict about objects at large distances, the Space Telescope offers astronomers an improved opportunity to determine which model is correct. Computer simulations of galaxy formation, based on the idea that cold dark matter dominates the density in the universe, can be tested by comparing the predictions of the simulations with the Space Telescope's observations of distant galaxies.

Also, because the Space Telescope is so versatile in its wavelength coverage and in the types of instruments available for its use, it will probably help cosmology in two other directions. The first improvement has to do with a series of specific areas of astronomy that need to be refined. One area concerns the value of the Hubble constant, which relates the distance of a galaxy to its velocity. The value of the Hubble constant must be known if we hope to calculate other key quantities such as the age of the universe and the value of the critical density of matter in the universe that separates the open and closed Friedmann models.

We can easily determine the recession velocity of any galaxy by measuring the Doppler shift in its spectrum with ground-based telescopes. But in order to obtain the value of the Hubble constant, we must know the distances of those objects for which the velocities have been measured, and that is a much more difficult task. In fact, this difficulty explains why our current best estimate of the Hubble constant, fifteen miles per second of velocity per million light-years of distance, is uncertain by about 30 percent; the constant's true value could lie anywhere in the range from ten to twenty miles per second per million light-years.

Can the Space Telescope help narrow our uncertainty as to the value of the Hubble constant? Astronomers believe that it will do so, by observing stars of a certain type, called Cepheid variables, at distances greater than those we can study well from the ground. Cepheid variable stars pulsate, that is, they expand and contract periodically, and are readily identifiable by their accompanying periodic changes in brightness. Using the now well-developed theory of the structure of stars, astronomers calculate that a close correlation should exist between each Cepheid's period of pulsation and the star's average luminosity. This general correlation between pulsation period and luminosity was confirmed nearly eighty years ago through Henrietta Leavitt's observations of Cepheid variable stars in the relatively nearby galaxy called the Large Magellanic Cloud.

To confirm that the luminosity of a star with a particular period of pulsation equals the period predicted by theory of pulsating stars, we can study Cepheid variables located within star clusters whose distance we know. Using the inverse-square law of apparent brightness, which relates any object's apparent brightness to its intrinsic luminosity, astronomers have verified that the pulsation theory is correct. Hence a Cepheid variable star furnishes astronomers with a "standard candle"—an object whose luminosity is known. Thus, wherever a Cepheid can be recognized, astronomers can calculate its distance by measuring its apparent brightness and comparing that brightness with the star's intrinsic luminosity, calculated from theory.

Cepheid variables offer excellent assistance to astronomers in determining the value of the Hubble constant. The way to use these stars is as follows: Astronomers search for stars whose light varies periodically, identify them as Cepheids, and determine the period of this variation. They know the luminosity of each Cepheid variable star because the theory predicts that a Cepheid with a particular pulsation period must have a particular luminosity. Using the inverse-square brightness law, they can then calculate the Cepheid variable's distance, and thus the distance of any galaxy that includes the Cepheid. They can then derive the value of the Hubble constant by dividing the velocity of that galaxy by its distance.

The difficulty with this approach, ever since Hubble first employed it, is that the velocities of relatively nearby galaxies arise less from the expansion of the universe than from random motions within the cluster of which they form a part. To obtain the true velocity arising from the universal expansion, we must observe all the bright galaxies in a cluster of galaxies, then average their velocities to obtain the recession velocity of the cluster as a whole.

This has been done quite accurately for the Virgo cluster of galaxies (Figure 64), whose recession velocity is about 600 miles per second. Unfortunately, the Virgo cluster of galaxies is so distant (about forty million light-years) that the Cepheid variable stars in its member galaxies are too faint to be found with ground-based telescopes. For now, astronomers must use other, less reliable standard candles to find the galaxies' distances. The value of the Hubble constant derived by using these standard candles is accordingly uncertain.

But because it can observe fainter objects, the Space Telescope will be able to find the individual Cepheid vari-

Figure 64: The Virgo cluster of galaxies, at a distance of about thirty million light-years. *National Optical Astronomy Observatories*

able stars in the Virgo cluster and to measure their apparent brightnesses. A relatively large amount of observing time could reasonably be allocated on the Space Telescope to astronomers who wish to carry out this project. As a result of this effort, astronomers will derive the Hubble constant more accurately, perhaps to an accuracy of 10 percent rather than 30 percent.

This increase in accuracy will pay dividends in calculations related to the large-scale structure of the universe. In particular, we shall know the distances to *all* distant galaxies more reliably, since astronomers determine those distances by measuring their velocities of recession and dividing those velocities by the Hubble constant (since $v = H \times d$, $d = v \div H$). As a result, astronomers will be able to calculate the age of the universe and the density of matter in the universe more accurately than we can now. When we compare this age and density with the ages of stars and the density of matter in galaxies, the latter quantities should enable us to home in on the correct cosmological model by eliminating incorrect ones.

LYMAN-α CLOUDS AND OTHER MYSTERIES

A second area of cosmology where the Space Telescope will play a cleanup role concerns the astronomically famous "Lyman-α clouds." The Lyman-α clouds, not to be confused with the Lyman-α galaxies discussed earlier, are clouds of gas between the galaxies. Lyman-α galaxies *emit* huge amounts of energy in ultraviolet radiation at the wavelength of Lyman-α ; Lyman-α clouds *absorb* radiation at this wavelength. The Lyman-α clouds were detected through their sharp absorption features (absorption that is highly circumscribed in wavelength), which astronomers have observed in the spectra of quasars with ground-based telescopes.

Quasars seem to be young galaxies with tremendous luminous activity in their centers; they are not completely

understood, but for the present discussion we can simply
regard them as bright, extremely distant points of light. As
the light from quasars passes through the billions of light-
years of intergalactic space that separate them from us,
occasionally—about every hundred million light-years or
so—the light encounters a cloud of hydrogen atoms, which
absorb some of the radiation from the quasar in a narrow
wavelength band in the ultraviolet region of the spectrum.
As shown in Figure 65, each cloud produces a Lyman-α

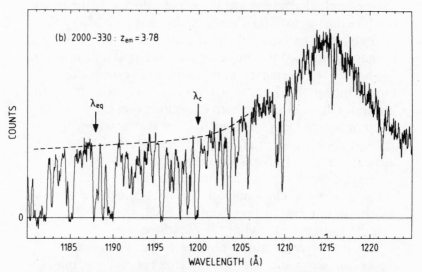

Figure 65: The ultraviolet region of the spectrum of the quasar 2000-
330, one of the most distant objects known. Each dip is a Lyman-α
absorption line originating in a Lyman-α cloud along the line of sight.
*Anglo-Australian Telescope, courtesy of R. W. Hunstead and the Cambridge
University Press*

absorption feature at a distinctive spectral wavelength.
The wavelength of the observed absorption is less than the
wavelength of the Lyman-α emission in the quasar but is
longer than the wavelength of the atmospheric cutoff of
ultraviolet radiation. Apparently the situation that we
observe is this: Clouds of hydrogen atoms lie along the line
of sight to each quasar. Each cloud is receding from us at a
velocity equal to its distance times Hubble's constant. The

wavelength of the absorption produced by each cloud depends on its velocity of recession.

We have discovered the Lyman-α clouds not by the radiation that they emit, but by the radiation that they block from quasars behind them. This leaves the nature of the clouds a mystery. Astronomers have examined the sky near the directions of the target quasars, hoping to see some emission from the intervening clouds. So far, they have had little luck, so we don't know the true nature of the clouds. The clouds could be located within yet-to-be-discovered galaxies moving at the same velocity as the galaxies, like the interstellar clouds within our own Milky Way, or they could be isolated clouds, unrelated to any galaxy, perhaps immersed in a more tenuous intergalactic medium.

The Space Telescope should help us to resolve this mystery. Unlike the Lyman-α clouds observable from the ground, which must have an enormous recession velocity in order for the Doppler effect to increase the wavelength of the Lyman-α absorption beyond the cutoff of the Earth's atmosphere, the Lyman-α clouds observable by the Space Telescope can have any amount of Doppler shift. Some of them may prove close enough for astronomers to observe any luminous matter they contain. Such observations will help us to figure out what the clouds are.

The list of mysterious objects to be observed by the Space Telescope includes the quasars themselves, intergalactic absorbing matter, star clusters near distant galaxies, gravitational lenses, and colliding galaxies. In dozens of ways, the sharpness of the images acquired by the Space Telescope and its ability to analyze ultraviolet light will make it a prime tool for research in cosmology and may shed new light on each of these types of objects.

The Space Telescope may revolutionize cosmology in another sense: by discovering new phenomena that we can hardly imagine now. The history of astronomy teaches us that each time a qualitatively new capability is brought into play, new discoveries follow. Galileo's telescope found sunspots, craters on the moon, the satellites of Jupiter, and the fact that the "Milky Way" consists of stars. The great refracting telescopes of the eighteenth and nineteenth centuries discovered proof that the planets orbit the sun, made the first measurement of the distance to a star, found interstellar gas, and first saw white dwarf stars. The giant reflectors of the twentieth century determined the ages of stars and their chemical composition, revealed the expansion of the universe, and made the discovery of quasars and of active galactic nuclei.

Now, with the Space Telescope probing the ultraviolet spectrum and producing images of exquisite sharpness of objects much fainter than ever seen before, we hope for more discoveries, some of which may clarify our understanding of the universe as a whole.

9

THE GREAT
OBSERVATORIES

THE SPACE TELESCOPE provides us with a commanding
view of the ultraviolet, visible, and near-infrared emission
of planets, stars, and galaxies. These objects emit other
types of radiation: far-infrared radiation, x-rays, and
gamma rays. To observe such radiation, we need space-
borne observatories other than the Space Telescope. Astronomers hope that the success of the Space Telescope
will mark the start of a new era in space-based observations of the cosmos, for they have either plans or actual
projects to build orbiting observatories to cover all the
currently unobservable regions of the spectrum.

Why do astronomers need these complex, expensive
observatories in orbit? Consider, for example, what
happens when stars have just formed through the collapse
of a particularly dense mass of interstellar gas. These
newborn stars remain immersed and hidden inside their
interstellar cocoons, as we observe in the Orion Giant
Molecular Cloud, which is located about 1,300 light-years
from us in the constellation Orion (Figure 66). Until the
1960s, when astronomers could make the first survey of
the sky with detectors of infrared radiation, no one even
knew that this giant cloud existed. But those infrared
detectors revealed that the familiar Orion Nebula is surrounded by the much larger Orion Giant Molecular Cloud.
Within this mass of interstellar gas and dust there lies a
region, invisible in ordinary light, that emits more than

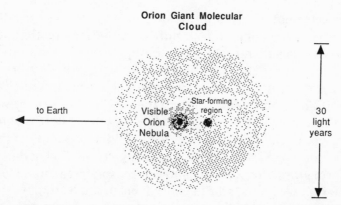

Figure 66: The giant molecular cloud that surrounds the visible Orion Nebula also includes a dense cloud in which stars are now forming; this cloud is best observed by means of the infrared radiation it emits. *Drawing by Marjorie Baird Garlin*

100,000 times as much energy as the sun does, all of it in the long-wavelength part of the spectrum. This region is invisible without an infrared detector.

What produces so much energy? The energy source for the infrared-emitting region is a cluster of massive young stars, each so hot that it emits mostly invisible ultraviolet radiation. You might guess that the ultraviolet radiation would make the young cluster a good target for the Space Telescope. But this is not so, as the ultraviolet radiation is completely blocked from our view by vast numbers of tiny dust particles in the cloud of gas from which the stars formed, making the region invisible even with the Space Telescope (as well as with visible-light telescopes on the ground). But the energy from the young stars must go somewhere. As the dust particles absorb the star's energy, they heat up, just as a road surface heats up on a summer day. Once hot, the dust particles start to emit radiation, but because they are only moderately hot, they emit their energy in the infrared part of the spectrum.

The net effect of the absorption and heating is that nearly all the stars' radiation is converted from ultraviolet into infrared emission, which was detected during the 1960s and can be readily studied using telescopes in orbit.

Star formation is a still-mysterious process that greatly interests astronomers. Computer simulations of the motions of gases in interstellar space imply that our own star, the sun, formed about 4.6 billion years ago, probably as the result of the gravitational contraction of an interstellar cloud. The planets now orbiting the sun, including our own Earth, were "debris" left behind during that fateful event. For sentimental reasons alone, we would like to understand how and when interstellar clouds of gas and dust collapse to form stars. But we have another reason to be interested: As far as we know, star formation occurs in a similar way throughout every galaxy in the universe. So if we hope to piece together the story of galaxies—how they formed and how they came to have their present configuration—we must also understand the process of star formation.

The Space Telescope will help. Its high-resolution spectrograph, used to analyze the ultraviolet radiation from hot stars that have already emerged from the dust clouds in which they were born, will find hundreds of sharp absorption lines in the stars' spectra. We know from studies with the *Copernicus* ultraviolet satellite that these absorption lines are produced by numerous small clouds of interstellar gas and dust that happen to lie along our line of sight— clouds too rarefied to collapse but composed of the same material as the denser clouds that do contract. The Space Telescope will therefore reveal the exact composition of matter slated to become stars at some time in the future. By implication, it will tell us the composition of the clouds that did collapse to form stars.

THE SPACE INFRARED TELESCOPE FACILITY

To study star formation directly, astronomers need an infrared telescope in space that will analyze the radiation from clouds of gas and dust in the process of collapsing to form new stars. NASA plans to launch such an instrument, the Space Infrared Telescope Facility (SIRTF), in 1998 (Figure 67).

Figure 67: The SIRTF satellite, which NASA hopes to launch for in-frared observations of the cosmos in 1998. *Courtesy of M. Werner, NASA, Ames Research Center*

The next few years should bring the start of SIRTF construction. SIRTF will have a mirror with a diameter of nearly one meter (thirty-nine inches). An essential, unique feature of SIRTF is that the telescope will be continuously cooled to an incredibly low temperature, just six degrees Fahrenheit above absolute zero. This cooling will eliminate any infrared radiation from the telescope, which would otherwise overwhelm the infrared emission from astro-nomical objects under observation. To cool the telescope and its mirror to such a low temperature is immensely difficult. The basic technique is to surround the telescope with liquid helium, which has an extremely low tempera-ture.

With SIRTF, astronomers can observe any objects that emit infrared radiation, including planets, stars, and galax-

Figure 68: The great infrared observatory of 1983 was the IRAS satellite. *NASA*

ies, with a sensitivity 1,000 times greater than SIRTF's predecessor, the Infrared Astronomy Satellite (IRAS), which flew in 1983 and made observations for nearly a year before its helium evaporated (Figure 68). SIRTF will have such great sensitivity to infrared that astronomers can divide the incoming infrared radiation into many separate wavelength "channels," or colors. This will enable them to identify various types of atoms and molecules responsible for the infrared emission, because each type emits at a different wavelength.

In the case of star-forming clouds, astronomers will be able to sort out the various types of molecules in the collapsing cloud, and even to obtain a rough measure (using the Doppler effect) of their motions. SIRTF will be so sensitive to infrared that it will be able to detect star formation in galaxies located in the most distant reaches of the universe. This exciting prospect means that we shall be able to observe star formation as it occurred billions of years in the past, in analogy to what must have been happening long ago in our own galaxy.

This trick of looking into the past is actually simple and unavoidable, because of the enormous amounts of time that have elapsed as light has traveled to us from the remotest parts of the universe. By studying galaxies at different distances, we can observe similar galaxies at what amounts to different stages in their evolution and can thus recapitulate the entire history of star formation from the time when galaxies formed billions of years ago to the present.

X-RAY ASTRONOMY FROM SPACE

X-rays are also of intense interest to astronomers. Until 1962, we had no idea that the high-energy photons that we call x-rays are being emitted by *any* object in the universe beyond the sun, which emits x-rays very weakly, but which we can detect as an x-ray source because the sun is relatively close to the Earth. In 1962, an x-ray detector that scanned the sky after being launched on a rocket that rose briefly above the atmosphere found evidence for a single pointlike source of x-rays in the constellation Cygnus. This source was later designated Cygnus X-1 as the first x-ray source found in that constellation. The rocket flight also revealed a diffuse "glow" of x-ray emission, apparently covering the entire sky, which astronomers now call the x-ray background.

In 1973, studies of Cygnus X-1 made with the orbiting x-ray observatory called *Uhuru* revealed that this x-ray source has the same position on the sky as a massive, luminous star easily visible to telescopes on Earth. However, it is not the *star* that emits the x-rays (because even luminous stars emit rather little of this high-energy radiation), but rather a hidden companion in orbit around the star. Study of the visible star's orbit allows us to conclude that the invisible object is massive, with at least ten times the mass of the sun.

The invisible companion is stealing gas from the star we see; this gas goes into orbit around the companion and slowly spirals inward. Because the companion is highly compressed, the intense gravitational field near its surface draws the gas into tight spiral orbits in which the gas reaches speeds up to one-tenth the speed of light. Friction within the gas moving at such high speeds heats it to temperatures of over ten million degrees, hot enough to emit x-rays. All of these concepts represent astronomers' conclusions—a bit of sweet detective work that relies on the observed properties and motions of the visible star.

Astronomers' interest in Cygnus X-1 is intensified by the theoretical prediction that any compact object with a mass greater than ten solar masses must be a black hole, a bizarre type of object predicted by Einstein's general theory of relativity but never before revealed by observation. Evidently, Cygnus X-1 began life as two ordinary stars orbiting one another. The more massive of the two stars evolved more rapidly and ended its stellar lifetime in a supernova explosion in which its core collapsed to a point, referred to as a black hole, unable to withstand the overwhelming crush of its own gravity. Later, as it grew older, the less massive star—the one that we can now observe— began to lose gas from its surface in the form of a "stellar wind." Some of this gas is now being captured into orbit by its black-hole companion. As strange as this interpretation may seem, it is the only one that makes sense to most astronomers.

BLACK HOLES

What is a black hole? How can a star as massive as ten suns occupy a single point and have no definite physical extension at all? This situation seems to fly in the face of intuition, as well as all that physics has taught us about the atomic composition of matter. This apparent contradiction arises because our intuition, and even our laboratory experiments, are useless when confronted with the enormous gravitational forces that sometimes arise in astronomical situations.

Once the core of a star attains a large enough mass and small enough size, the equations of physics tell us that no force—not even the pressure of an extremely hot gas or of atomic nuclei in contact—is adequate for the core to resist the inward crush of its own gravitation. Then a collapse without limit is inevitable. The collapse proceeds to the point that even light waves attempting to escape from close to the newly formed black hole cannot do so, because the gravitational force on them (or, if you prefer, the gravitational bending of space) is so large. The failure of light to escape gives the black holes their fearsome name.

Bizarre as it is, Cygnus X-1 is not the only place where a black hole exists in the universe; several similar objects have been found by x-ray astronomers. But even more

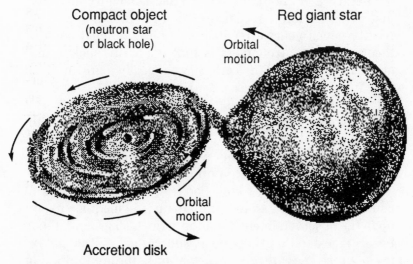

Figure 69: The accretion disk around a compact object is made of gas that spirals into the object as it orbits a companion red giant star. The disk is continually replenished by gas from the companion. *Drawing by Marjorie Baird Garlin*

numerous than black holes are binary star systems in which the compressed companion has not collapsed so much that it has formed a pointlike black hole, but instead has formed an extremely compact object called a neutron star. In these objects, each as massive as a star but only about twenty miles across, the pressure exerted by rapidly moving particles called neutrons is barely sufficient to prevent further gravitational contraction. Like black holes, neutron stars in binary systems can steal gas from their companions. This theft provides energy for x-ray sources, since the particles in the gas heat and collide as they spiral inward at increasing velocities toward the surface of the neutron star (Figure 69).

WHAT CREATES THE X-RAY BACKGROUND?

The diffuse x-ray background discovered by the first x-ray rocket is evenly distributed over all parts of the sky. The spectrum of this radiation resembles that of a gas at a temperature of about 700 million degrees Fahrenheit, but just where such hot gas exists remains unexplained twenty-seven years after its detection. The most popular theory among astronomers is that the x-ray background arises from the combined effect of hot gas in many millions of individual x-ray sources, distributed through the remotest reaches of observable space, where the galaxies' x-ray images are so closely packed that they can mimic a smooth, diffuse glow.

There are some problems with this theory, however. If the emitting objects are hypothesized to be normal galaxies, then the total x-ray emission from the black holes and neutron-star binary stars within them is inadequate to add up to the observed intensity of the x-ray background. It has been suggested that these x-ray sources are distant galaxies in which the rate of star formation is far higher than in our own galaxy, with the result that massive stars and x-ray emitting binary systems are more numerous there. If so, then the Doppler effect from the expansion of the universe would have shifted the spectrum of the individual stellar objects to longer wavelengths and lower energies, contrary to observation. Perhaps the x-ray-emitting objects are quasars, that is, galaxies whose enormous luminosities suggest the presence of a single supermassive black hole, with up to one billion solar masses, at the center of each quasar. Quasars do, in fact, emit copious amounts of x-rays, but their x-ray spectra don't appear to be similar to that of the x-ray background.

HOT GAS IN THE UNIVERSE

Some astronomers have suggested that the diffuse x-ray background arises from hot gas that is not concentrated around black holes in galaxies or around supermassive black holes in quasars, but rather spread thinly and uniformly throughout the entire universe. With such a theory, astronomers can explain why the observed emission is smoothly distributed over the sky and why the spectrum has the observed spectral distribution. But this theory has its own problems. One problem is hypothesizing the energy source for the enormous amount of heat that would be stored in the hot gas.

The only source that is energetic enough seems to be "superconducting cosmic strings," proposed to exist by particle physicists but never observed. These incredible theoretical objects, flashing through space at close to the speed of light, could carry enormous electrical currents in a filament of primordial matter with less than one thousand-trillionth the diameter of an elementary particle. With such high currents and enormous speeds, cosmic strings (if they exist) would emit super-intense beams of electromagnetic radiation, which would heat the diffuse gas in the universe to the temperatures required to explain the x-ray background.

The diffuse x-ray background remains a mystery, and the physics of black holes is almost as mysterious. To

Figure 70: The AXAF x-ray satellite. Unlike the primary mirror of the Space Telescope, the AXAF mirrors are located at the front of the telescope, where they can gently guide the x-rays to a focus at the rear. For comparison, note the size of the astronaut at the lower left. *NASA*

understand the universe we live in, it seems imperative to learn more about both. For that reason, astronomers place a high priority on creating NASA's proposed Advanced X-Ray Astrophysics Facility (AXAF), an x-ray telescope with a mirror forty-five inches in diameter, configured so that incoming x-rays reflect from its mirror at low angles of incidence, thereby being brought to a common focus (Figure 70). X-rays will not reflect from mirrors if they arrive from directions nearly perpendicular to the mirror surface; instead, they will be absorbed by the mirror.

AXAF will be a true imaging telescope, capable of producing *pictures* of objects in their x-ray emission. This is a tricky thing to do with x-rays. Fortunately, the principles of the AXAF telescope have been proven in flight by the *Einstein* x-ray observatory that NASA operated from 1978 to 1981. *Einstein* studied many black-hole and neutron-star sources, as well as extremely hot gas in clusters of galaxies and even moderately hot gas in the atmospheres of normal stars like our sun. AXAF, some 100 times more powerful than *Einstein*, will enable us to detect x-ray galaxies at the limits of the observable universe, and to sort out once and for all what types of objects produce the diffuse x-ray background. Whether the answer turns out to be black holes, or diffuse hot gas heated by cosmic strings, or (perhaps most likely) something entirely unexpected, we shall learn basic new facts about the universe.

GAMMA-RAY ASTRONOMY: THE LAST FRONTIER?

Beyond x-rays in energy are the gamma rays, photons of the highest energies, so energetic that special techniques are necessary to capture them. Earlier observations have revealed that the sky glows uniformly with gamma rays, providing astronomers with another mystery to solve. Nearer to the solar system, gamma rays arise within interstellar clouds when energetic particles in space—the "cosmic rays"—collide with atoms in the interstellar gas. In fact, astronomers can determine the total mass within an interstellar cloud simply by measuring the number of gamma rays produced in it by these collisions. More mass implies more collisions and more gamma rays.

Gamma rays also originate in collapsed objects such as the neutron stars and black holes in x-ray binary systems, as well as near supermassive black holes in quasars. They also arrive from "gamma-ray bursters," objects that emit intense, short bursts of gamma rays but that appear to have no counterpart emission in other regions of the spectrum of electromagnetic radiation. One gamma-ray burster lies in the direction of the remnants of a supernova that exploded in a nearby galaxy, which suggests that the gamma rays have somehow been generated by a black hole or neutron star left behind by the supernova explosion. However, the energy in the bursts is so great that no theory has yet explained them to astronomers' complete satisfaction.

All theories of how and why some stars explode as supernovas predict that radioactive nuclei, such as nickel 56 and cobalt 56, are formed by the intense energy of the explosion. These nuclei decay into nuclei of iron 56 and emit gamma rays as part of the decay process. In February 1987, a supernova exploded in the relatively nearby galaxy known as the Large Magellanic Cloud, the first time in three centuries that a supernova explosion was close enough to be seen with the unaided eye. During 1987 and 1988, every available astronomical instrument in the Southern Hemisphere and in space was directed, at least for a time, toward the supernova. This included a gamma-ray detector on board the *Solar Maximum Mission* satellite, which was designed to study energetic events on the sun.

By 1988, the matter ejected from the supernova had thinned out to the point that gamma rays could escape from the explosion, and in fact the *Solar Maximum Mission*'s detectors found them, confirming astronomers' calculations that the nuclei of heavy elements such as nickel, cobalt, and iron are fused in the fiery furnaces of supernovas. This fact has great importance from a cosmic perspective, because such heavy elements are observed throughout the Milky Way Galaxy and were included in the solar system during the formation of the sun and its planets. On Earth, these heavy elements, which form the basis for our technical civilization and indeed are essential to life itself, came from stars that exploded billions of years ago, before the sun, moon, and Earth began to form.

Gamma-ray astronomy offers us a unique opportunity to study the synthesis of heavy elements in space, as well as to improve our understanding of black holes, neutron stars, and the objects that produce gamma-ray bursts, whatever they may be. For this reason, NASA is constructing a Gamma-Ray Observatory, to be launched in late 1990, soon after the Space Telescope (Figure 71).

Taken as a unified set of satellites, the Space Telescope, Gamma-Ray Observatory, Advanced X-Ray Astrophysics

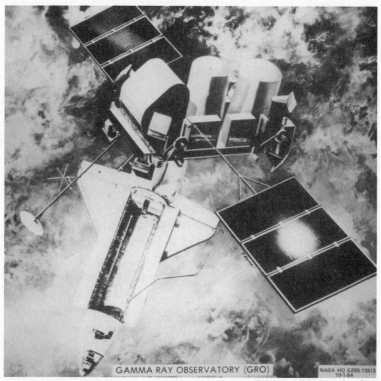

GAMMA RAY OBSERVATORY (GRO) NASA HQ EZ85-126(3) 10-1-84

Figure 71: An artist's conception of the Gamma-Ray Observatory (GRO) to be launched by the Space Shuttle in 1990. *NASA*

Facility, and Space Infrared Telescope Facility will form a suite of powerful orbiting observatories to cover the entire electromagnetic spectrum from gamma rays to infrared emission. This range spans a factor of about *ten billion* in photon wavelength. These long-lived observatories' unprecedented power to carry out astronomical science in space qualifies them for the title "Great Observatories," a designation once reserved for the giant mountaintop facilities, whose reach the space-borne instruments will complement and extend. The future of much of astronomy lies in space, free from atmospheric blockage that until now has robbed us of the chance to see the universe in gamma rays, x-rays, ultraviolet, and infrared radiation.

10

THE SPACE TELESCOPE
AND THE FUTURE OF
ASTRONOMY

ASTRONOMY TODAY CONCERNS itself with the nature, origins, motions, and evolution of the matter in the universe. Astronomy has always done so. When Democritus, at the end of the fifth century B.C., speculated on the atomic nature of matter, and Aristarchus, in the middle of the third century B.C., speculated that the Earth moves around the sun, these Greek philosophers were trying to understand the material universe in the same spirit that we attempt to embrace today. Galileo (1564–1642) was equally at home with a telescope and with laboratory apparatus; he saw material objects in motion in the heavens at the same time that he tried to deduce the laws of motion on Earth.

Isaac Newton (1642–1727) fused together these strands of celestial and terrestrial motion with his theories of gravitation and dynamics. He also invented a new kind of mathematics, called calculus, to help calculate the precise relationships among time, space, and motion. Such bold advances set the stage for the spectacular rise of science during the past two centuries, both as a sustained intellectual effort and as the practical basis for industrial civilization.

The revolution of understanding that Newton initiated has still not run its course. On the one hand, the study of electromagnetic radiation, of atoms, of atomic nuclei, and most recently of elementary particles has brought us to a better understanding of the submicroscopic world, a world in which Werner Heisenberg's uncertainty principle reigns and forces physicists to think in terms not of certainties but of probabilities. On the other hand, Albert Einstein's theories made physicists realize that Newton's laws of motion and of gravitation only approximate a more exact theory, the general theory of relativity, which offers a coherent description of the entire universe, including the motion of matter at nearly the speed of light. We now stand at the opening of an era in which, as in the age of Newton, a new synthesis of our theories of matter and of the entire universe will become possible and even necessary.

A MODERN SYNTHESIS

Examples of this growing synthesis abound. Consider the supernova explosion observed on February 23, 1987, in the Large Magellanic Cloud, 160,000 light-years from the Earth. Astronomers used the supernova's light to determine its position and its distance. Physicists, examining the data from particle detectors buried deep in the Earth (Figure 72), discovered a brief pulse of neutrinos, elemen-

Figure 72: A neutrino detector containing 5,000 tons of water is located half a mile beneath Ohio. This device detected neutrinos from Supernova 1987A in the Large Magellanic Cloud. *Department of Energy*

tary particles that are extremely difficult to detect, from the explosion.

Astrophysicists had predicted that neutrinos would carry away most of the energy released in a catastrophic supernova explosion. For twenty years nuclear physicists had been confident that exploding stars do produce neutrinos, but had never detected them. Then, from the first supernova that occurred close enough to Earth and after a sensitive enough detector had been constructed, they found the neutrinos, just nineteen of them detected out of trillions upon trillions that are believed to have passed through the Earth (roughly one million through each human being). The numbers of neutrinos, the times of detection, and the energies of the observed neutrinos all agreed with physicists' predictions.

The few seconds that saw the detection of the neutrino pulse from the supernova also saw the unification of theories of the enormously large and the incredibly small: a star with twenty times the mass of the sun was behaving in the manner specified by the properties of an elementary particle that carries only one one-hundredth the energy of a proton. Deep-space neutrino astronomy was born, as physicists and astronomers made another step toward understanding the universe on the basis of the properties of elementary particles.

The unification of our theoretical understanding of the large and the small has been paralleled by a synergy in our experimental techniques. The development of microwave electronics during World War II allowed the construction of the particle accelerators and the radio telescopes of the postwar period, and the development of solid-state electronics has made possible sensitive detectors for the elementary particles produced in accelerators and the photons that astronomers detect from deepest space. In addition, progress in physics and astronomy has been greatly accelerated by the development of ever more powerful computers.

All of this progress coalesces in the science of cosmology, the study of the evolution of the entire universe. In the big-bang model of the universe, the temperature was so great when the universe was less than a trillionth of a second old that in order to understand the physics that governed what went on then, we must learn more about elementary particles at the very highest energies. This means constructing enormous particle accelerators. The quest to understand how the universe evolved from those earliest moments onward drives astronomers to look further back in time, hence farther out in space, as it drives physicists to study the behavior of particles at ever-higher energies. The quest therefore requires both larger telescopes and bigger accelerators.

Many obstacles still stand in the way, at various levels, of a comprehensive description of the universe. First, astronomers are already beginning to encounter astronomical costs for new telescopes. And particle physicists are finding that earthquakes, because they can wrench one part of a building away from another, impose limits on the physical size of their accelerators. So, from a very practical standpoint, we may face terrestrial limits on our quest.

On a more sweeping level, a universal factor inhibits astronomers from investigating directly what occurred during and just after the big bang itself. The event must have been shrouded in hot gas that absorbed the radiation from the big bang. Within a few seconds after the big bang, the universe was filled with gamma rays. However, those gamma rays were soon absorbed by matter, and the radiation that replaced them was less energetic than the gamma rays, because of the cooling that took place as the universe expanded.

By now, 10–15 billion years after the big bang, the radiation has been degraded in wavelength by the Doppler effect to the point that it appears in the millimeter-wavelength and microwave parts of the spectrum, where it can be observed by radio telescopes as the cosmic background radiation. The fact that this radiation exists with a certain intensity tells us a great deal about the early universe, but any trace of the variations from place to place in the intensity of this background radiation has been wiped out by the rapid shuffling back and forth of photons as the universe has expanded. The situation resembles that within a star, where we cannot learn details about the center of the star from observations of its surface. Therefore, when physicists propose new theories about the behavior of matter in the early universe, astronomers cannot test them directly. Hence progress in this area will be slow.

What astronomers can and will do, by using the Space Telescope and the other space-borne Great Observatories, is to look as far as possible into the history of the universe. The events that we hope to observe occurred hundreds of millions of years—not seconds, as we might wish—after the big bang. Even so, they come from a time so much earlier than the present that the traces of still earlier events may yet be left upon them.

FORMATION OF GALAXIES

An example discussed in Chapter 9 concerns the formation of galaxies. Amazingly, the structures that we now observe as galaxies may be condensations of matter in an otherwise expanding universe that were initiated by small disturbances, much as raindrops form in a rising and cooling column of air. In the case of raindrops, these initial disturbances could be the dust particles on which water could condense; in the case of galaxies, the initial disturbances could be bumps in the quantum fields present in the early universe, bumps that gathered matter around them through gravitation.

Current theories in particle physics actually predict the stage in the expansion of the universe at which galaxies should begin to form (when the universe was about one billion years old), as well as the stage at which the formation of galaxies should reach a crescendo (when the universe was about 1.6 billion years old). This is a prediction that astronomers can sink their teeth into, and it is exciting to learn that a few of the most distant galaxies that have been observed *do* appear to be in the process of formation at about the time expected.

The Space Telescope is ideally suited to observe faraway objects that we see not as they are but as they were many billion years ago. The Space Telescope can obtain sharp images of these young galaxies, and can obtain ultraviolet spectra to help reveal what the galaxies consist of. Other instruments will be brought to bear as well. Large ground-based telescopes, because of their large collecting areas, will allow finer discrimination in analyzing the galaxies' visible-light spectral features. Radio telescopes may detect the emission from gas located in the young galaxies; this is important because—at least in our own galaxy—stars form only where gas is abundant in space.

The Space Infrared Telescope Facility (SIRTF) may reveal more about these distant, faint, young galaxies. Once again, from studies of our own galaxy, we know that star-forming regions glow brightly in the infrared, owing to the intense ultraviolet light from luminous young stars that interstellar dust grains absorb and then reprocess into infrared radiation.

Young galaxies can also be detected by the Advanced X-Ray Astrophysics Facility (AXAF). The intense heating of interstellar gas as it condenses to form the galaxy should lead to x-ray emission from the hot gas. Also, fast-evolving stars that explode as supernovas should blast "shock waves" that heat the gas still further.

The study of individual young galaxies will certainly become a key frontier in astronomy during the next decade. In addition, studies of the *distribution* of young galaxies in space will come to the fore to test theories based on the particle physics supposed to dominate the early universe. Such theories predict that any structure in the distribution of galaxies should develop only slowly as the universe ages and as the gravitational forces among galaxies draw them closer together to form clusters. No doubt, astronomers will develop methods to map the distribution of far more galaxies than can be studied individually, and to do this for faint, distant, and therefore young galaxies as well as for the nearer, older galaxies that we already have observed. It will be fascinating to check whether galaxies indeed become less clustered as one examines more distant, hence younger, galaxies, as is predicted by theory.

Behind these investigations lurks the gravitation exerted by the dark matter, invisible to any observatory on the ground or in space. Evidence is accumulating that such matter exists, but its nature is unknown. Is it abundant enough to make the geometry of the universe flat, as particle physicists predict? On the level of submicroscopic physics, can the strange particles making up the dark matter—if they exist—somehow be detected by Earth-based experiments? Theory predicts that these particles should be spread through space everywhere, including our galaxy and our solar system. Detecting them near the Earth would constitute a triumph for theoretical and experimental physics and would forge a link between earthbound physics and the astronomical observations of the dark matter that apparently controls the fate of the universe.

For now these plans remain dreams, but they are the dreams that motivate young scientists and inspire them to make new discoveries. The foresight of our scientific leaders, of government officials, and of the public in providing magnificent instruments such as the Space Telescope to pursue the quest for a unified understanding of the universe is remarkable—all the more so in view of the pressing demands on the governmental budget. As astronomers, we are proud to be part of this venture, and we are pleased to invite the public to participate in the excitement of the Space Telescope, the instrument that the public decided should be built to place our eyes above the atmosphere.

GLOSSARY

Absolute temperature scale - Temperature measured on a scale that begins at absolute zero and increases by the same units as those used in the centigrade (Celsius) system, so that water freezes at 273.16 degrees and boils at 373.16 degrees.

Absolute zero - The lowest point on any temperature scale, the temperature at which all motion ceases (except for certain quantum mechanical effects). Absolute zero occurs at −273.16 degrees on the centigrade (Celsius) scale and at −459.67 degrees on the Fahrenheit scale.

Absorption - The blocking or elimination of electromagnetic radiation.

Absorption line - A region of the electromagnetic spectrum, limited in frequency and wavelength, from which electromagnetic radiation has been removed, so that the intensity of the radiation falls below that of the neighboring spectral regions.

Advanced X-ray Astrophysics Facility (AXAF) - NASA's planned orbiting observatory to study x radiation from space.

Angular momentum - A measure of the amount of spin possessed by an object, determined by the mass of the object times its rate of spin times the square of its size in the direction perpendicular to the axis of spin.

Angular size - The apparent size of an object, measured by the angle the object subtends, in degrees of arc, minutes of arc, and seconds of arc.

Atom - The smallest unit of a chemical element, consisting of a nucleus of one or more protons and neutrons surrounded by one or more electrons in orbit around the nucleus.

AURA - The Association of Universities for Research in Astronomy, a consortium of twenty universities that operates several astronomical facilities for the National Science Foundation and that administers the Space Telescope Science Institute for NASA.

Black hole - An object with such enormous gravitational force that nothing, not even electromagnetic radiation, can escape from it.

Black-hole radius - The critical radius of any object, equal to three kilometers (two miles) times the object's mass in units of the sun's mass. An object that contracts within its black-hole radius will become a black hole, and nothing will be able to escape from inside the black-hole radius.

Blueshift - A shift to shorter wavelengths and higher frequencies caused by the Doppler effect in an approaching object.

Blurring - The spreading out of waves of visible light or of other forms of electromagnetic radiation, which prevents the observer from obtaining as clear a view as would otherwise be possible. Typically blurring arises in the Earth's atmosphere, as each portion of the atmosphere acts as a lens and diverts individual beams of light from what would otherwise be straight-line trajectories.

CCD - A charge-coupled device, an electronic detector of electromagnetic radiation, based on silicon chips that respond to incoming electromagnetic radiation by producing an electric current, and which can detect electromagnetic radiation with great sensitivity. Each of the eight CCDs used in the Space Telescope's wide-field/planetary camera has more than 640,000 individual picture elements, called pixels.

Celestial sphere - The imaginary sphere of the sky, centered on the Earth, and on which the sun, moon, planets, stars, and galaxies may be visualized as all at the same distance from the Earth.

Celsius temperature scale - Centigrade temperature scale.

Center of mass - The point within an object or group of objects that makes the quantity of mass times distance from that point the same on either side of that point, in any direction.

Centigrade temperature scale - The scale of temperature that registers the freezing point of water as 0 degrees and the boiling point as 100 degrees.

Comet - A fragment of primitive solar-system material, made of ice and dust, typically a few kilometers across, that moves in a huge, elongated orbit around the sun.

Copernicus satellite - An ultraviolet-detecting satellite launched by the United States in 1972, which continued to send observational data to Earth until 1981.

Corona - The outermost parts of the sun (and of other stars), millions of miles in extent, consisting of highly rarefied gas heated to temperatures of millions of degrees.

Coronal hole - A region of the solar corona of relatively low density, through which the particles that form the solar wind can easily escape.

Coronal loop - A region above the solar surface where the sun's magnetic field arches into the solar corona and then returns to the surface, often connecting cooler surface regions, called sunspots.

Cosmology - The study of the universe as a whole, and of its structure and evolution.

Degree of arc - A unit of angular measure equal to $\frac{1}{360}$ of a full circle.

Diffraction grating - A mirror ruled with thousands of closely spaced parallel lines. Radiation reflected from such a mirror is spread into its constituent wavelengths and frequencies.

Doppler effect - The observed change in the frequency and wavelength of electromagnetic radiation that arrives from a source in relative motion toward or away from the observer. The amount of the change caused by the Doppler effect increases in proportion to the relative velocity. Approach velocities increase the frequency and decrease the wavelength; recession velocities decrease the frequency and increase the wavelength.

Einstein satellite - An x-ray-detecting satellite, the first to obtain good images of cosmic x-ray sources, sent into orbit above the Earth's atmosphere in 1978.

Electromagnetic forces - One of the four basic types of forces, acting between particles with electric charge, either as repulsive forces (between particles with the same sign of charge) or as attractive forces (between particles with opposite signs of charge).

Electromagnetic radiation - Waves of energy, consisting of massless particles called photons, characterized by a frequency (number of vibrations per second) and a wavelength (distance between successive wave crests).

Electron - An elementary particle with one unit of negative electric charge and a mass that is only $\frac{1}{1,836}$ of the mass of a proton.

Element - The set of all atoms that have the same chemical properties and the same number of protons in each atomic nucleus.

Elementary particle - A fundamental particle of nature, of which the most prominent examples are quarks, electrons, and photons.

Elliptical galaxy - A galaxy with an ellipsoidal distribution of its stars, hence a galaxy whose shape on a photograph is that of an ellipse.

ESA - The European Space Agency, whose thirteen members are Austria, Belgium, Denmark, France, West Germany, Ireland, Italy, the Netherlands, Norway, Spain, Sweden, Switzerland, and the United Kingdom with Finland an associate member. ESA developed the faint-object camera aboard the Space Telescope.

Fahrenheit temperature scale - The temperature scale that registers the freezing point of water as 32 degrees and the boiling point as 212 degrees.

Faint-object camera (FOC) - The camera aboard the Space Telescope provided by ESA, designed to exploit the full power of the Space Telescope by observing the faintest possible objects with the greatest possible resolving power, both in ultraviolet and in visible light.

Faint-object spectrograph (FOS) - The spectrograph aboard the Space Telescope used to study faint objects, using two Digicon systems of detectors, one sensitive to reddish visible light and one sensitive to bluish visible light and ultraviolet radiation.

Far infrared - The long-wavelength part of the infrared region of the electromagnetic spectrum.

FGS - Fine guidance sensor.

Filter wheels - Aboard the Space Telescope, wheels that hold forty-eight different filters, each of which removes electromagnetic radiation at particular frequencies and wavelengths from the beam of incoming radiation. As a result, through combinations of filters, the Space Telescope, using the CCDs in its wide-field/planetary camera, can determine the amount of radiation of different frequencies and wavelengths that reaches the CCD detectors.

Fine Guidance Optical Control Sensors - Fine guidance sensors.

Fine guidance sensors (FGSs) - Devices sensitive to ultraviolet and visible light, used aboard the Space Telescope to detect whether or not guide stars are in the telescope's field of view, and thus to direct the telescope accurately toward a particular direction.

Fixed-head star trackers - Small telescopes with a wide field of view aboard the Space Telescope, used to find relatively bright stars to serve as preliminary guide stars, in order for the fine guidance sensors to track the actual, fainter guide stars.

FOC - Faint object camera.

FOS - Faint object spectrograph.

Frequency - The number of vibrations per second, usually measured in units of hertz (one oscillation per second).

Galaxy - A large group of stars, typically numbering in the billions, held together by their mutual gravitational attraction, typically about a hundred thousand light-years in diameter.

Gamma-Ray Observatory (GRO) - NASA's planned orbiting observatory, to be launched in 1990 to observe gamma rays from cosmic objects.

Gamma rays - Electromagnetic radiation with the smallest wave-lengths and highest frequencies of all types.

Geosynchronous orbit - A synchronous satellite orbit around the Earth.

Globular star cluster - A spherical or ellipsoidal group of stars, usually ten or twenty light-years in diameter and containing several hundred thousand, or a few million, individual stars.

Gravitational forces - One of the four basic types of forces (the other are electromagnetic, strong, and weak). Gravity is always attractive. For any two particles, the amount of gravitational force varies in proportion to the product of the two masses, and in inverse proportion to the square of the distance between the objects' centers.

Great Red Spot - A semipermanent feature in the upper atmosphere of Jupiter, apparently a sort of cyclone, several times larger than the Earth.

Greenhouse effect - The trapping of infrared radiation by a planet's atmosphere, which raises the temperature of the planet's surface above the value it would have if the infrared radiation could escape directly into space from the surface.

GSFC - The Goddard Space Flight Center, a NASA center in Greenbelt, Maryland, which issues the commands to guide the operation of the Space Telescope.

HEAO - NASA's High Energy Astronomical Observatory satellites, of which two were successfully orbited during the 1970s.

High-resolution spectrograph (HRS) - One of the two spectrographs aboard the Space Telescope, used for observations of relatively luminous objects and especially useful for measuring the exact wavelengths and frequencies of ultraviolet and visible light.

High-speed photometer (HSP) - The sensitive photometer (light-measuring device) aboard the Space Telescope, capable of detecting both ultraviolet and visible light, and of distinguishing events in time separated by only one hundred-thousandth of a second.

HRS - High-resolution spectrograph.

HSP - High-speed photometer.

HST - Hubble Space Telescope.

Hubble's law - The rule that summarizes our observations of the expanding universe: the recession velocities of the galaxy clusters that we observe equals a constant, called Hubble's constant, times the clusters' distances from us.

Hubble Space Telescope - NASA's automated telescope in space, a reflecting telescope with a primary mirror ninety-four inches in diameter, capable of pointing with 0.012 second of arc accuracy and of observing in the ultraviolet and visible-light regions of the electromagnetic spectrum.

Image intensifier - A device similar to the light-sensitive cathode-ray tube in a television camera, and which responds to a single electron produced by the impact of electromagnetic radiation by producing thousands of times as many electrons, intensifying a faint image into a brighter image when the many electrons strike a surface that glows upon electron collision.

Infrared - Electromagnetic radiation with slightly longer wavelengths and slightly smaller frequencies than those of visible light.

Interstellar medium - In a galaxy, matter spread diffusely among the stars, typically consisting of gas (made of small atoms and molecules) and dust (particles with a few billion atoms each, still of microscopic size).

Ion - An atom that has lost one or more of its electrons.

Ionization - The process by which ions are produced, typically by collisions with atoms or electrons, or by interaction with electromagnetic radiation.

IRAS - The *Infrared Astronomy Satellite*, which observed the universe in infrared radiation during 1983.

Isotope - The subset of a given element that consists of those atoms whose nuclei all contain the same number of neutrons, as well as the same number of protons.

IUE - The *International Ultraviolet Explorer*, a satellite that was launched into synchronous orbit in 1978 and continues to make observations in the ultraviolet.

Kilometer - A unit of distance equal to 0.62137 mile.

Koesters prism - A prism used in the fine guidance sensors of the Space Telescope to detect accurately the exact location of a star whose light reaches the prism.

Light-year - The distance that light travels in a year, equal to 9.46 trillion kilometers or about 6 trillion miles.

Local Group - The small cluster of about twenty galaxies to which our Milky Way galaxy belongs.

Magnetic field - A field of force in space, created by a magnet or by an electric current, that guides the trajectories of electrically charged particles by exerting an electromagnetic force.

Milky Way - The galaxy to which our solar system belongs, whose central regions appear as a band of light or "milky way" in the night sky seen from Earth.

Minute of arc - One-sixtieth of a degree of arc.

Molecule - A stable grouping of two or more atoms, bound together by electromagnetic forces among the electrons and nuclei in the atoms.

MSFC - NASA's Marshall Space Flight Center in Huntsville, Alabama.

NASA - The National Aeronautics and Space Administration, created by the U.S. government in 1958 and charged with the peaceful exploration of space.

Near infrared - The portion of the infrared region of the electro-magnetic spectrum that lies relatively close to the visible-light region of the spectrum.

Nebula - A diffuse mass of interstellar matter, often lit from within by young, hot stars that have recently formed from the interstellar gas and dust.

Neutron - A subatomic particle with no electric charge, one of the two basic constituents of an atomic nucleus.

Neutron star - A tremendously dense object, typically about a dozen kilometers across, formed from the central core of a col-lapsed star, in which almost all of the particles are neutrons.

Nuclear fusion - The joining of two nuclei under the influence of strong nuclear forces, which typically reduces the total mass and increases the total kinetic energy that the particles contain.

Nucleus (plural: **nuclei**) - (1) The central concentration of matter in an atom, composed of one or more protons and zero or more neutrons. (2) The central concentration of matter in a galaxy, typically spanning no more than a few light-years in diameter.

OAO - One of two NASA satellites called Orbiting Astronomical Observatories, capable of ultraviolet observations of the uni-verse, launched in 1968 and 1972. The second successful OAO was renamed the *Copernicus* satellite.

OSO - One of NASA's Orbiting Solar Observatories, launched during the early 1970s, which studied the sun in visible and ultraviolet light.

Ozone - Molecules made of three oxygen atoms each, which, high in the Earth's atmosphere, shield the Earth's surface against ultraviolet radiation by absorbing it.

Parallax effect - The apparent displacement in position of stars caused by the Earth's yearly motion in orbit around the sun.

Parsec - An astronomical unit of distance equal to 3.26 light-years, or approximately 19 trillion miles.

Photometer - An instrument that measures the intensity of light or other forms of electromagnetic radiation.

Pick-off mirrors - In the Space Telescope, one of four flat mirrors held by a rigid arm at an angle of forty-five degrees to the beam of incoming radiation, which diverts a small portion of that radiation either to one of the fine guidance sensors or to the wide-field/planetary camera.

Pixel - A single element in an image, corresponding to a single dot in a mosaic picture.

Planet - One of the nine largest objects in orbit around the sun; also, similar objects that may be in orbit around other stars.

Planetary nebula - A shell of gas surrounding an aging star, expelled from the star itself, that is heated and partially ionized by the radiation emitted from the hot surface of the star.

Primary mirror - The main mirror of a reflecting telescope, which gathers electromagnetic radiation and directs it toward a smaller secondary mirror, which in turn brings it to a focus.

Proper motion - A star's apparent motion against the background of much more distant stars, after the parallax effect has been subtracted, which arises from the star's own velocity through space with respect to the sun.

Quark - An elementary particle, of which protons and neutrons are made. Quarks have not been observed in isolation, but physicists are nevertheless sure that they exist.

Quasar - A quasi-stellar object, almost starlike in appearance, but actually one of the most powerful and most distant sources of energy known to exist in the universe.

Radiation - *See* Electromagnetic radiation.

Radio - Electromagnetic radiation with the longest wavelengths and smallest frequencies of all types.

Reaction wheel - Aboard the Space Telescope, one of four rotating flywheels that are used to make the telescope rotate either more rapidly or less rapidly around a particular axis of rotation.

Red giant star - A star that has passed through the major portion of its life, during which it fused hydrogen nuclei (protons) into helium nuclei at a constant rate, and has now swelled and cooled its outer layers while its central, nuclear-fusing core has contracted and grown hotter.

Redshift - A shift to longer wavelengths and lower frequencies caused by the Doppler effect in a receding object.

Reflecting telescope - A telescope that employs a mirror as the primary means of collecting and focusing visible light or other forms of electromagnetic radiation.

Refracting telescope - A telescope that employs a lens as its primary means of collecting and focusing radiation.

Resolving power - Of a telescope, the ability to reveal two closely separated sources of radiation as individual sources rather than as a single, combined source. Hence, the ability to obtain a sharp view of the universe.

Secondary mirror - In a reflecting telescope, a small mirror mounted in the beam of radiation that strikes the primary mirror. Radiation is directed by the primary mirror onto the secondary, which (in the Space Telescope) reflects the light through a small hole in the center of the primary mirror toward instruments that detect and analyze the radiation.

Seyfert galaxy - A spiral galaxy with an unusually bright, compact nucleus.

Solar activity cycle - The cyclical change, repeating every eleven years, in the strength of the sun's magnetic field, the number of sunspots, and the number of particles emitted in the solar wind.

Solar maximum - The time of maximum output of particles and radiation in the solar activity cycle.

Solar system - The sun and the objects in orbit around it, which include nine planets, fifty-four known satellites of the planets, thousands of smaller objects called asteroids, and billions of comets.

Solar wind - Streams of charged particles expelled from the sun, which expand past the Earth and the other planets as a stream of electrons and ions, the latter typically nuclei of hydrogen, helium, oxygen, carbon, nitrogen, and neon with one or more electrons in orbit around each nucleus.

Space Infrared Telescope Facility (SIRTF) - NASA's planned infrared observatory, to orbit outside the Earth's atmosphere, with its mirror cooled to three degrees above absolute zero, which should achieve great sensitivity in observing infrared radiation from cosmic objects.

Space Shuttle - The manned vehicle used by the United States to place payloads into orbit. It consists of a manned orbiter powered by liquid-fuel rocket engines, an external tank for liquid fuel, and two solid-fuel rockets. The shuttle reaches an orbit several hundred miles above the Earth's surface.

Space Telescope - Hubble Space Telescope.

Space Telescope Science Institute - The international research facility at the Homewood Campus of the Johns Hopkins University, from which the scientific research with the Space Telescope is directed and where the data returned from the Space Telescope may be analyzed. It is managed by AURA for NASA.

Spectrograph - An instrument that spreads electromagnetic radiation into its component frequencies and wavelengths for detailed examination at each individual frequency and wavelength.

Spectroscopy - The observation and analysis of the spectra of radiation from distant objects.

Spectrum (plural: **spectra**) - The distribution of electromagnetic radiation over its component frequencies and wavelengths.

Spiral galaxy - A galaxy characterized by a disk of stars, within which the youngest, brightest stars appear within spiral patterns called spiral arms.

Star - A self-luminous mass of gas held together by its own gravitation, at the center of which energy is released by nuclear fusion.

Star cluster - A group of stars born at almost the same time and place, capable of remaining a unit for billions of years because of the mutual gravitational attraction of its stars.

Stratoscope - One of a series of balloon-borne telescopes, conceived by astronomers at Princeton University, which were carried to altitudes of about twenty miles during the late 1950s and early 1960s, and which secured images of the sun, planets, and nearby star systems with a sharpness previously unobtainable.

Strong forces - Forces that act between protons and neutrons, binding them together into atomic nuclei, and which are effective only at distances of 10^{-13} centimeter or less.

Sunspot - A region on the sun's surface that is somewhat cooler and darker than the surrounding surface.

Supermassive black hole - A black hole with millions or billions of times the sun's mass.

Supernova - An exploding star, visible for weeks or months, even at enormous distances, because of its tremendous energy output.

Synchronous orbit - An orbit around a planet (the Earth in particular) at an altitude where a satellite moves at just the speed at which the planet rotates; hence, an orbit in which an orbiting satellite remains nearly stationary above a particular point on the planet.

TDRSS (pronounced "teedress") - NASA's Tracking and Data Relay Satellite System, a system of satellites in high Earth orbit, used to relay data from the Space Telescope to Earth and to send commands from Earth to the Space Telescope.

Temperature - The measure of the average energy of random motion within a group of particles.

Uhuru satellite - An x-ray-detecting satellite launched into a synchronous orbit from a floating platform at the Earth's equator off Kenya in 1971.

Ultraviolet - Electromagnetic radiation with frequencies somewhat greater and wavelengths somewhat less than those of visible light.

Visible light - Electromagnetic radiation with intermediate frequencies and wavelengths, to which human eyes are sensitive.

Wavelength - The distance between two successive wave crests in a wave such as an electromagnetic wave.

White dwarf - A star that has completed the fusion of helium nuclei into carbon nuclei in its interior, and thereafter slowly radiates away the energy stored while it was still performing nuclear fusion. It is typically a few thousand kilometers in diameter and of much lower luminosity than an active star.

Wide-Field/Planetary Camera - Camera aboard the Space Telescope capable of gathering twenty million bits of information per picture, with a spectral sensitivity that ranges through the entire ultraviolet and visible-light regions into the near infrared.

X-rays - Electromagnetic radiation with somewhat greater frequencies and smaller wavelengths than those of ultraviolet radiation.

Zenith - The point on the celestial sphere directly overhead as seen by a particular observer.

FURTHER READING

Bahcall, John, and Lyman Spitzer, Jr. "The Space Telescope." *Scientific American* 247:1 (July 1982), p. 40.

Clark, David. *The Cosmos from Space.* New York: Crown Publishers, Inc., 1987.

Field, George, and Eric Chaisson. *The Invisible Universe.* Boston: Birkhauser Books, 1985.

Ghitelman, David. *The Space Telescope.* New York: Gallery Books, 1987.

Harwit, Martin. *Cosmic Discovery*, 2d ed. New York: Basic Books, 1989.

Longair, Malcolm. "The Scientific Challenge of Space Telescope." *Sky & Telescope* 69:4 (April 1985), p. 306.

McDougall, Walter. *The Heavens and the Earth.* New York: Basic Books, 1985.

Tucker, Wallace. "The Space Telescope Science Institute." *Sky & Telescope* 69:4 (April 1985), p. 295.

Tucker, Wallace, and Karen Tucker. *The Cosmic Inquirers.* Cambridge: Harvard University Press, 1986.

INDEX